THE
TEACHING OF PHYSICS

ENGLAND:	BUTTERWORTH & CO. (PUBLISHERS) LTD. LONDON: 88 Kingsway, W.C.2
AUSTRALIA:	BUTTERWORTH & CO. (AUSTRALIA) LTD. SYDNEY: 20 Loftus Street MELBOURNE: 473 Bourke Street BRISBANE: 240 Queen Street
CANADA:	BUTTERWORTH & CO. (CANADA) LTD. TORONTO: 1367 Danforth Avenue, 6
NEW ZEALAND:	BUTTERWORTH & CO. (NEW ZEALAND) LTD. WELLINGTON: 49/51 Ballance Street AUCKLAND: 35 High Street
SOUTH AFRICA:	BUTTERWORTH & CO. (SOUTH AFRICA) LTD. DURBAN: 33/35 Beach Grove
U.S.A.:	BUTTERWORTH INC. WASHINGTON D.C. 20014: 7300 Pearl Street

THE TEACHING
OF PHYSICS

J. W. WARREN. M.Sc., Ph.D.

*Senior Lecturer, Physics Dept.,
Brunel College, London*

LONDON
BUTTERWORTHS
1965

Suggested U.D.C. Number: 53 : 371·3

| *First Published* | - | - | - | - | - | 1965 |
| *Reprinted* | - | - | - | - | - | 1966 |

©

J. W. Warren

1965

Printed in Great Britain by
The Lewes Press Wightman & Co. Ltd., Friars Walk, Lewes, Sussex.

CONTENTS

	PAGE
Preface	ix

Chapter 1. Forces   1
Newton's laws of motion. The third law. Circular motion. Weight and weightlessness. The deformation of bodies. Intermolecular forces.

Chapter 2. Gravitation   10
The law of Newton. The law of tradition.

Chapter 3. Energy   13
Kinetic energy. The mass-energy relationship. Nuclear energy.

Chapter 4. Molecular Theory   18
Interactions and motions of molecules. Elasticity. Surface energy and latent heat. Thermal expansion and electrical resistance. The purpose of the kinetic theory of gases. Properties of molecules. The effects of the walls. Intermolecular forces in different phases. The equation of Van der Waals. Absolute zero.

Chapter 5. Temperature   31
Scales of temperature. The ideal gas scale. Thermodynamic scales. Traditional treatment of temperature scales. The expansion of gases. The law of Charles. Conclusion.

Chapter 6. Thermodynamics   42
Thermodynamic equilibrium. Reversible and irreversible changes. Expansion processes. Expansion of thermally insulated substances. Expansion of gases with thermal insulation. The principal specific heats. Entropy.

CONTENTS

PAGE

Chapter 7. Heat Transfer   53
Thermal Conductivity. Convective heat transfer. Verification of Newton's law of cooling. Problems in convective heat transfer.

Chapter 8. Electromagnetic Radiation   59
Heat and light. Short-wavelength radiations. Optical spectra.

Chapter 9. Principles of Current Electricity   65
Potential difference and electromotive force. Resistance and Ohm's law. Verification of Ohm's law. Effective and R.M.S. values. Inductance. The magnetic circuit.

Chapter 10. C.G.S. Units of Electricity and Magnetism   76
Introduction. System I. System II. Comments upon the two systems.

Chapter 11. Atomic Physics   81
Radioactivity. Absorption of radiation in matter. Radiation hazards. Atomic structure. The traditional treatment of atomic structure. The Bohr theory. Some basic measurements.

Chapter 12. Miscellaneous   91
Elastic hysteresis. Rolling friction. The human eye. The leaf electroscope. Measurement of low resistances. Semiconductors. Photoelectric emission. Signs of increments.

Chapter 13. Definitions   97
General. Mass. Specific heat. Resistivity. Equilibrium vapour pressure. Discussion.

Chapter 14. The Unreality of G.C.E. Physics   104
Physics as an algebraic exercise. Applicability of laws. The electromagnetic Doppler effect. Invented results. Materials. Obvious errors.

Chapter 15. Ancient and Modern 112
Advancing science. Calorimetry. Photometry. D.C. Electrical measurements. Atomic Physics. Gamma for gases.

Chapter 16. Summary and Conclusions 119
Errors of fact. Errors of reason. Criticism. Current reforms. Conclusion.

Appendix 124
Notes and References.

PREFACE

Every academic subject is taught in accordance with a tradition which embodies certain accepted facts and methods of thought. To a very large extent the individual teacher or examiner, however learned or independently minded, is constrained to teach or examine in conformity with the tradition of his subject. Examinations and textbooks interact upon each other. The writers of textbooks must attempt to cover the syllabuses of the principal examinations and to give their readers the type of information which is acceptable to the examiners. Examiners can generally only ask those questions which can be answered on the basis of currently available texts.

Over a period of many years there has developed a traditional interpretation of elementary physics for teaching purposes. Somehow it has happened that into this tradition has been incorporated much that is false in fact or absurd in logic. It is the purpose of this book to discuss a number of these mistaken ideas. The tradition of physics teaching is most unsound at sixth-form level, hence most of our discussion is concerned with material commonly taught for A-level and S-level G.C.E., although both more elementary and more advanced ideas are included.

In asserting that a particular idea is traditionally accepted it is not intended to imply that every single teacher, author or examiner accepts it. Generally the erroneous ideas which are condemned in this book are those which the author has found to be taught to a large proportion of those who study the subject.

In deciding what is the traditional belief on any issue the sources consulted have been as follows:

Current textbooks, particularly those written for sixth-form physics courses.

The examination papers of G.C.E., National Certificate and similar examinations.

Popular accounts of science such as are commonly read by students.

University and college students whom the author has taught over many years.

Colleagues and friends of the author in various branches of the teaching profession.

Correspondence and conversations with examiners and authors.

This book is intended to be a criticism of a profession, not of individuals. For this reason no passages have been quoted verbatim from any source, but instead paraphrases of typical statements in textbooks have been given to exemplify the ideas which are commonly taught.

I am very grateful to Professor C. A. Hogarth and Mr. R. Hall who read the manuscript and made several valuable suggestions. The facts and opinions expressed in the book are, however, entirely the responsibility of the author.

Chapter 1

FORCES

Newton's Laws of Motion

Dynamics is one of the most essential foundations of physics. This subject was first soundly established in the 17th century on the basis of the axioms usually known as Newton's laws of motion. Although certain modifications are now known to be necessary for the precise formulation of dynamics in accordance with modern developments these laws are sufficiently exact for nearly all applications. A clear understanding of Newtonian dynamics is of the first importance to every student of science and engineering.

First Law—A body at rest remains at rest, and a body in motion moves with uniform speed in a straight line, unless a force acts on it.

Second Law—The rate of change of momentum of a body is proportional to the impressed force and takes place in the direction of that force. By suitable choice of units one may say that the rate of change of momentum *is equal* to the force.

Third Law—If a force acts on a body then an equal and opposite force must act upon another body.

In these laws 'force' means the resultant of all externally applied forces. Momentum is the product of mass times velocity, thus for a body of constant mass, rate of change of momentum is equal to mass times acceleration. Hence force is often wrongly defined as mass times acceleration. As mass can vary in several circumstances this is unsound and misleading.

These laws are fundamentally linked with the law of gravitation. The planets and satellites are the most familiar objects whose motions can be studied with precision, and which are remote from other bodies. They do not move in straight lines with uniform speed, hence if we accept the first law we must suppose that they are acted upon by those forces which we call gravitational.

Newton's laws of motion are not obvious, indeed they conflict with any ordinary interpretation of familiar phenomena. Common experience suggests that it is necessary to exert a force to maintain uniform motion, whilst it is natural to distinguish between states of 'rest' and 'uniform motion in a straight line'. For these reasons the imagination repudiates the laws so that we find, nearly three centuries after Newton's work, the formal statements and definitions of qualified physicists are often pre-Newtonian in spirit. For example, force is often defined as 'that which tends to move or deform a body'. Such a definition clearly violates the first law, which asserts that bodies move in the absence of forces, and the second law which states that the effect of a force is a change of momentum. Similarly statics is often defined as the science of bodies at rest and dynamics as the science of bodies in motion. Such a distinction is meaningless in the light of the first law. In fact, statics is the science of bodies in mechanical equilibrium, that is systems of *unaccelerated* bodies, whilst dynamics is the study of accelerated bodies. Such questions of terminology are important in themselves, but they are above all symptoms of the grip which pre-Newtonian ideas has upon our minds, as will be shown in the following paragraphs.

The Third Law

This law is nearly always taught in one of the epigrammatic forms 'action and reaction are equal and opposite' or 'to every action there is an equal and opposite reaction'. Such phrases are in themselves meaningless and it is therefore not surprising that the law is very generally misunderstood. Different forces acting on the same body are often described as action and reaction forces, hence the Third Law becomes confused with the totally distinct question of equilibrium.

A typical textbook illustration of this law is as follows. 'When a man is standing still he is at rest under the action of two forces, his weight acting downwards and an equal and opposite reaction from the floor upwards'. Apart from a completely false application of Newton's third law we may note the misleading introduction of the idea of rest. There is further difficulty in the meaning of the word *weight* and in the fact that the two forces acting on the man are not actually equal in magnitude or opposite in direction (p. 5). It is very usual for students to be taught similarly that a body lying on a table is acted upon by equal and opposite action and reaction forces. If the table collapses it is supposed that the 'reaction' ceases to exist, thus breaking the law which is not generally appreciated to be inviolable within the range of classical ideas.

It frequently happens that when one force is applied to a body other forces automatically arise acting upon the same body and opposing the original force. In popular speech such an opposing force is the 'reaction' to the original cause, but it is certainly not the reaction in Newton's sense, which has nothing to do with cause and effect. Confusion on this point can probably only be prevented by eliminating altogether from physics teaching such useless epigrams as 'action and reaction are equal and opposite'.

Circular Motion

It is easily shown that if a body moves with constant speed V in an arc of radius of curvature R then it has an acceleration towards the centre of curvature equal to V^2/R. If the body has a constant mass M there is a rate of change of momentum directed towards the centre equal to MV^2/R. By Newton's second law this must be caused by a force, directed towards the centre. This centripetal force is, with suitable choice of units, equal to MV^2/R. This extremely important result is so simple in form, and is so easily derived from the Second Law, that there does not seem to be any excuse for its not being fully understood. Unfortunately it is the subject of endless confusion.

In very many elementary courses and in popular presentations of physics, as well as in many courses in engineering and mathematics, an imaginary 'centrifugal force' is introduced into this problem without any explanation. This force, which is obviously considered to be quite real, is supposed to be equal and opposite to the centripetal force, thereby giving a zero resultant. Thus there is a body which is being accelerated but which is said to be under the action of no force. This absurd analysis is very commonly given for the cases of orbital motion of planets and electrons, and for the whirling of bodies at the ends of strings. It is even more confused when applied to vehicles turning corners. A diagram typical of those given in several textbooks is shown in *Figure 1*. The vehicle turning a corner on a banked track is shown under the action of forces Mg downwards and the 'centrifugal force' MV^2/R horizontally, giving a resultant directed into the earth. It is apparently the action of this

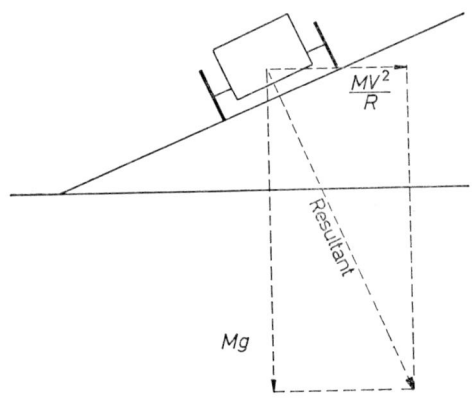

FIGURE 1.

resultant which is supposed to cause acceleration in the horizontal plane in the opposite direction to the centrifugal force. Similar treatment is given in the case of a cyclist leaning over on turning on level ground.

It seems inconceivable that any rational person who has been taught rules as explicit as Newton's first and second laws could fail to appreciate immediately that these analyses of circular motion are in complete disagreement with them. Nevertheless, hundreds of teachers and authors either do not notice the inconsistency, or do not care.

These absurdities in the teaching of mechanics may have

a number of causes. One is the confusion concerning the third law discussed in the previous section. There is known to be a centripetal force acting on the circulating body, and the 'reaction' is mistakenly assumed to act on the same body. Another explanation may lie in the naïve interpretation of everyday experience, for one can easily believe one is experiencing an outward force when in a cornering vehicle. Finally, it is possible that the ideas may arise from vague memories of the principle of d'Alembert. Whatever may be the cause, this is a totally unjustified burden to lay upon our students.

Weight and Weightlessness

The weight of a body is usually defined as the force of gravity acting upon it. It is also commonly taught that weight is measured by a spring balance. These two ideas are not quite consistent since the earth is rotating and a body 'at rest' on the surface of the earth is not in equilibrium but has an acceleration towards the axis. For this reason the force applied to the spring balance is slightly different in magnitude and direction from the gravitational attraction of the earth on the suspended body. The vector difference is the force required to keep the body in circular motion.

Weight is also normally identified with Mg, where M is the mass and g is said to be the gravitational acceleration. The precise meaning of g is seldom explicitly stated, but it seems that usually it represents the *apparent* acceleration, that is the acceleration of a freely falling body as measured by an observer fixed at a point on the surface of the earth. The force of gravity on the body is however the product of M times the acceleration towards the centre of the earth. As this is not the same as g in magnitude and direction Mg is not equal to the weight as usually defined.

Some authors define weight as the force exerted by a supported body upon its support. This is of course the force measured by a spring balance, and is equal to Mg. It is important to note that according to this approach an unsupported body has no weight.

The differences between the various meanings given to weight may seem unimportant, since the radial acceleration of a point on the equator is only about three thousandths of the magnitude of g; but this is a very precisely measurable quantity, and the difference is very significant. Further, we are concerned here with basic concepts of great importance for the proper understanding of mechanics, and it is essential to be clear in our definitions. It is particularly important to be clear upon which body weight acts. Ambiguity upon this point may be partly responsible for some of the confusion concerning action and reaction forces discussed previously.

The extent of the confusion which can arise in this field is shown by the difficulties which students experience with the popular concept of the 'weightlessness' of a space traveller.

A passenger in an artificial satellite is subject to certain physiological effects attributed to 'weightlessness'. What meaning is to be attributed to this term? The magnitude of the earth's gravitational attraction upon the spacetraveller is not much less than at ground level. If weight is this gravitational attraction he certainly still has weight. Sad to relate, the majority of the 'space age' students questioned by the author suppose that the spacemen have been shot so high that they are effectively beyond the reach of the earth's gravity. These students cannot, of course, explain how it is that the satellite follows a curved course under the action of no force. The same difficulty arises for those other students who believe that weightlessness is the result of the gravitational attraction being opposed by centrifugal force. This latter absurdity is only applicable to circular orbits, hence it is now taught that weightlessness is characteristic of circular motion. Actually the physiological effects will be the same whenever one is unsupported, whatever the shape of the orbit.

Men and other land creatures are normally subjected to gravity acting throughout the volume of the body, and supporting forces applied to limited areas of the surface. All physiological processes affected by mechanical causes must be adjusted to these circumstances. When travelling in a space vehicle without the rockets working the supporting forces are removed and 'weightlessness' results. Clearly we should only use the term weightlessness if by weight we understand the force exerted on or by the support.

Other physiological effects are suffered by travellers in fast aircraft when these craft undergo large accelerations. Such effects are commonly attributed to the acceleration of the person. However, acceleration by itself cannot have any such effects. The astronaut in a freely-falling space vehicle is in exactly the same physiological condition as if he were in zero gravitational field or if he were being accelerated by the far greater field of a massive planet. The 'black-outs' suffered by jet pilots are the result of their being subjected to large forces applied to external surfaces of their bodies.

The Deformation of Bodies

It is very usual to state that a body is stretched by a force, although the application of a single force to a body will produce an acceleration. Stretching is the result of a tension, that is a system of two equal and opposite forces applied in the same straight line to different parts of the body. It is important to note that the two forces which constitute a tension act on the same body and are not 'action and reaction'.

It is similarly usual to state that a body is twisted or sheared by a couple. Now it can be shown that a couple is the angular analogue of a force, that is the effect of a couple is to produce angular acceleration. To produce a shear it is necessary to use equal and opposite couples suitably applied. It is only too common to find a shear stress illustrated by such figures as 2a and 2b. The forces shown in *Figure 2a* would cause angular acceleration, whilst the force in *Figure 2b* would of course accelerate the specimen out of the figure.

Intermolecular Forces

The tradition of physics teaching neglects both physical principles and logical consistency. This is shown particularly clearly by the fantastic treatment of intermolecular forces which is almost universally accepted. This is considered in greater detail in Chapter 4, but will be illustrated here by the consideration of the 'explanations' of surface tension which are given in very many textbooks of heat and mechanics. These explanations all assume purely attractive forces between molecules, which decrease with increasing separation. Two typical accounts are as follows:

'A molecule well inside the liquid will be pulled equally in all directions by the other molecules. One in the surface will have an unbalanced force pulling it downwards at right angles to the surface'.

'When a molecule is at the surface it will not be attracted equally in all directions, for there is no liquid above it. In consequence such molecules will tend to move towards the interior of the liquid'.

According to Newton's second law, if the outer molecules are subjected to such unbalanced forces they must accelerate inwards. Hence according to these analyses all matter must be collapsing inwards. The fact that such a preposterous

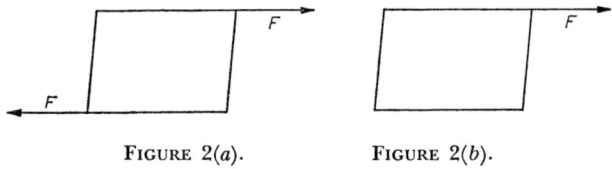

FIGURE 2(a).   FIGURE 2(b).

absurdity should have been taught in all seriousness as a scientific theory for at least a generation is staggering. As so many physicists have accepted this treatment without question it is justifiable to indicate two further weaknesses. Firstly, if a molecule were to be displaced slightly from its equilibrium position, according to the traditional model it will be attracted more strongly to the nearer neighbours and hence will accelerate *away from* the equilibrium position. Thus even the interior of the material will be totally unstable. Secondly, if we imagine any plane drawn through a drop of liquid, according to this model the two parts must accelerate towards each other.

A further absurdity in the description of surface phenomena is sometimes met in the explanation of angle of contact. A diagram like *Figure 3* is shown and it is stated that the three forces acting on a molecule in the solid-liquid-vapour interface are in equilibrium. It should be obvious that such

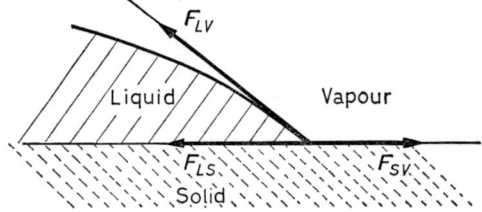

FIGURE 3.

a system of forces cannot possibly have a zero resultant, but this diagram has been published in textbooks for at least thirty years.

Chapter 2

GRAVITATION

The Law of Newton

Newton's law of gravitation can be stated in the form: Every particle of matter in the universe attracts each other particle by a force which is proportional to the product of their masses and inversely proportional to the square of their distances apart.

$$\text{Algebraically} \quad F = \frac{GM_1 M_2}{r^2}$$

where F is the attractive force between any two particles of masses M_1 and M_2 separated by a distance r, and G is the universal gravitational constant.

It is essential to note that this law applies to *particles*, that is bodies whose dimensions are infinitesimal compared with all other distances involved in the analysis. To apply the law to bodies of finite extension it is necessary to integrate. This can sometimes be done most easily by expressing the law in the form of Gauss's theorem. From this three important results can easily be established, namely:

(1) The external gravitational field of a rigid spherically symmetric body is the same as if the whole mass were concentrated at the centre. (A body is spherically symmetric if its properties at any point are dependent only on the distance of the point from the centre, and not on the polar angle. In the present case the only relevant property is density.)

(2) A rigid spherically symmetric body will be attracted as if its mass were concentrated at the centre, when in a gravitational field obeying the inverse-square law.

(3) Inside a uniform shell there is no gravitational field caused by the shell.

The results (1) and (2) permit the calculation of the interactions of the sun, the planets and the satellites. These bodies are very nearly spherically symmetric and hence it is a very good first approximation to regard their masses as concentrated at their centres, in calculations of orbital motion. In very precise calculations corrections have to be made for departure from perfect symmetry and it is possible, for example, to obtain information upon the shape of the earth by the study of the motion of artificial satellites. The result (1) was originally derived by Newton to justify the application of the inverse square law to the attraction of bodies close to the surface of the earth.

A *rigid* body has been specified above since non-rigid bodies will be distorted by gravitational fields. This distortion is exemplified by the tides.

The Law of Tradition

One common error in the statement of the inverse square law with application to bodies in general, instead of particles. The distance between two finite bodies is normally understood to mean the distances between their nearest points, hence this statement of the law predicts an infinite force between bodies in contact.

Very many teachers avoid this gross error only to perpetrate one almost as absurd. This is the expression of the law of gravitation in terms of the distance between the centres of mass of the bodies. Typically the law is stated as follows: 'Newton proved that any real body attracts another in exactly the same way as it would if the whole of its mass were concentrated in one point, called the centre of mass, or centre of gravity, of the body'. A student can easily test this statement using, for example, a saucer and a coin. By inspection the centre of mass of the saucer is above the surface near the middle. If a coin is placed near here and moved about, its centre of mass will pass very near to that of the saucer. Although gravity is normally very weak the force must become very large when the centres of mass are very close. The bodies should now oscillate until they come to rest with their centres of mass coincident. Now an infinite force is required to separate them! This misstatement of Newton's law leads to many similar absurdities, for example it predicts an infinite attractive force between an orange and its peel. The idea of a centre of gravity as a fixed point at which a single equivalent force can be applied is extremely useful, but it can only be employed in a straightforward way in those gravitational problems in which the field is uniform.

Sometimes it is taught that the inverse square law can be applied assuming the masses to be concentrated at the centres, *provided the bodies are uniform spheres.* Astronomical bodies such as the earth and sun are certainly not uniform, as their densities increase towards their centres. According to this approach then the law of gravitation cannot be applied to planetary and satellite motions, which are the very problems which it was introduced to solve. The restriction to uniform spheres is quite unnecessary as spherical symmetry is sufficient, and the proof is just as easy in this case as for uniform spheres.

The theorem that a uniform spherical shell produces no gravitational field inside itself is subject to various misinterpretations. Sometimes the shell is said to be hollow, but this is irrelevant. Sometimes it is even taught that there is no gravitational field inside a uniform hollow shell, which implies a gravitational screen. Probably this absurdity arises from a confusion with electrostatic shielding where also an inverse square law is applicable.

This theorem is often misapplied to calculations of the gravitational acceleration in mines. On the assumption that the earth has uniform density it can be shown that the internal field is proportional to the distance from the centre. On this basis it is taught that g has its maximum value at the surface and decreases on going below ground. Actually the mean density of surface rocks is less than half the mean density of the earth, from which it can be shown that g continues to increase on going down mines to all depths at present obtainable.

CHAPTER 3

ENERGY

Kinetic Energy

Kinetic energy is the energy which a body possesses because of its motion, and is therefore the work which the body will do whilst it is being brought to rest. This is sometimes expressed as 'the work done in bringing the body to rest'. This phrase is ambiguous as it leaves doubt as to whether work is being done by the body, or by the retarding force. This ambiguity may be responsible for the belief sometimes encountered that work must be done to stop a moving body. Examination candidates have been expected to show that a given amount of energy is sufficient to stop a certain moving body.

To obtain a formula for the kinetic energy of a body moving with speed V the work done by it whilst stopping in a distance x is expressed as

$$KE = \text{work} = \int_0^x (\text{force}) \, dx$$

It is usual to equate force to mass times acceleration, whence

$$KE = -M \int_0^x \left(\frac{dV}{dt}\right) dx$$

By simple changes of variable, using the fact that $V = dx/dt$ this gives

$$KE = \tfrac{1}{2} M V^2$$

This result is so commonly used that it very generally believed to be an exact and universally applicable formula, and is often supposed to be a definition of kinetic energy.

It is not generally appreciated that this formula is only applicable to bodies of constant mass.

The Mass-energy Relationship

It can be shown that if the laws of conservation of mass and momentum are to be consistent with the special theory of relativity the mass of a body must be a function of its speed. The mass M as measured by an observer with respect to whom the speed is V is given by

$$M = \frac{M_0}{\sqrt{(1 - V^2/c^2)}} \qquad \ldots \ldots (3.1)$$

where M_0 is the rest-mass, that is the value of M when V is zero, and c is the speed of electromagnetic radiation in a vacuum.

When considering bodies with speeds which are not negligible compared with c this variation of mass must be taken into account. In particular the exact form of Newton's second law must be used, the usual approximation that force is mass times acceleration being no longer valid.

To obtain an expression for the kinetic energy of a body of speed V we can calculate the work done in accelerating the body from rest to this speed in the x-direction.

We have $KE = \text{work done} = \text{force} \times \text{displacement}$. Thence since force is rate of change of momentum

$$KE = \int_0^x \frac{d}{dt}(MV)\,dx = \int_0^t V \frac{d}{dt}(MV)\,dt$$

$$= M_0 \int_0^t V \frac{d}{dt}\left\{\frac{V}{\sqrt{(1 - V^2/c^2)}}\right\} dt$$

Integrating by parts this gives

$$KE = M_0 \left[V \times \frac{V}{\sqrt{(1 - V^2/c^2)}} \right]_0^V - \int_0^V \frac{V\,dV}{\sqrt{(1 - V^2/c^2)}}$$

Whence
$$KE = (M - M_0)c^2 \qquad \ldots \ldots (3.2)$$

This is the exact formula for kinetic energy. If we substitute from equation (3.1) into equation (3.2) and expand the square root by the binomial theorem we find that the kinetic energy tends to $\frac{1}{2}M_0 V^2$ for values of V much less than c.

Students who believe that kinetic energy must always be $\frac{1}{2}MV^2$ sometimes suppose incorrectly that they can allow for relativistic effects by using this formula with the mass given by equation (3.1) instead of the rest-mass.

We see from equation (3.2) that the increase of the mass of a body as a result of motion can be written

Increase of mass = Kinetic energy $\div c^2$

This suggests a general relationship, namely that a quantity of energy E has a mass M given by

$$E = Mc^2 \qquad \ldots \ (3.3)$$

This is Einstein's mass-energy relationship. It has been illustrated above for kinetic energy and can be extended to potential energy by considering a head-on collision of two similar elastic particles. Before and after the collision the system consists only of two non-interacting bodies each of mass M, hence the total mass is $2M$. At the instant of impact the two particles are at rest, hence each has mass M_0. Now equation (3.1) was originally derived on the assumption that mass is conserved, hence the mass of the system must still be $2M$. What is there now in the system which can have the missing mass $2(M - M_0)$? If energy is conserved, at the moment of impact the system contains potential energy of elastic deformation equal to the kinetic energy before and after impact. This energy is then equal to $2(M - M_0)c^2$, so that if the relation (3.3) applies to potential as well as kinetic energy the conservation laws can be preserved.

It is important to be able to extend these ideas to particles of zero rest-mass such as photons and neutrinos. Since such particles carry energy they must have masses given by equation (3.3). This is consistent with equation (3.1) if they move with speed equal to c, since M_0 is then zero for finite M.

Nuclear Energy

The mass-energy relationship is very widely misunderstood. It is usual to teach that this relationship implies the conversion of mass into energy, neither quantity being conserved. It is supposed that the generation of nuclear power and the operation of the atomic bomb depend upon such 'mass-energy conversion' and that nuclear energy is in this respect distinct from other forms of energy. This is in conflict with relativity dynamics which makes no particular reference, implicit or explicit, to nuclear energy but refers throughout to energy in general.

The phrase 'conversion of mass into energy' suggests that the mass of a system may be reduced whilst the energy is increased, but the mass-energy relationship is derived on the assumption that both mass and energy are conserved. Consider the annihilation of an electron-positron pair. According to the conservation laws the mass of the annihilation radiation must be the sum of the masses of the particles before their interaction. What has happened is that the rest-mass of the particles, which is the mass of the energy of their static fields, is converted into the mass of photons, which is the mass of the energy of their alternating fields. Usually we measure the masses of particles rather than determining their energies directly, whilst we measure the energy of radiation rather than determining its mass directly, whence perhaps arises the idea that mass has been 'converted' into energy. The inadequacy of this point of view is apparent on considering the acceleration of a charged particle, in which case we can measure the increases of both its mass and its energy by essentially direct methods.

The mass-energy relationship does not distinguish in any way between different types of energy. If a power station delivers a million kilowatt-hours of energy this will have a mass of forty milligrams whatever the source of power. Thus for a coal-fired station if it were possible to measure exactly the mass of coal and air used and the mass of ash and fumes produced it would be found that the latter would weigh forty milligrams less than the former, the balance having been transmitted along the supply cables to increase the internal energy (and hence the mass) of material in the user's premises. The same will apply to the fuel and 'ash' of a nuclear generating station. A hydroelectric plant obtains energy from the gravitational field of the earth. The mass of the energy stored in this field is reduced as water flows to regions of lower potential energy. In all these cases correction must of course be made for waste energy. In no case would it be possible to determine directly the mass of the material used with the precision necessary to check the calculation. The mass-energy relationship does not 'explain' nuclear power any more or less than it 'explains' coal power. The great importance of the relationship in nuclear physics lies in the possibility of measuring the masses of nuclei with such precision that in some cases it is possible to predict the energies which may be evolved or absorbed in nuclear interactions, and in other cases it is possible to use measured energies to determine masses. Unless we wish to discuss such calculations it is irrelevant and misleading to introduce mass-energy into our teaching on nuclear power. The recent introduction of these sophisticated ideas into an O-level syllabus can only be deplored.

Many teachers feel that they must use the mass-energy relationship to explain the much greater energies available from nuclear fuels and explosives as compared with chemical ones. Since the relationship also applies to chemical energy such 'explanations' mean nothing. The great difference is explicable only in terms of atomic structure. Chemical energy is derived from small changes in the outermost parts of the interacting atoms. Nuclear energy is derived from proportionally larger changes in the structure of the nucleus, which contains all but about one part in four thousand of the mass of a typical atom.

Chapter 4

MOLECULAR THEORY

Interactions and Motions of Molecules

By *molecules* we understand the constituent particles of matter, which are usually ions but are sometimes atoms or chemical molecules.

Familiar properties of condensed matter show that there must be forces of interaction between molecules, which must vary with the distance apart. Solid and liquid materials at a given temperature normally have very constant densities. They resist uniform compression by forces which increase steadily with increasing strain to indefinitely high values. Clearly then at short distances the intermolecular forces are repulsive, becoming stronger as the distance becomes smaller. Solids, and also liquids under favourable conditions, also strongly resist extension up to a limiting strain at which rupture occurs. Hence the intermolecular forces are attractive at medium distances, increasing in strength up to a limiting value, then falling off with further increase of distance. The fact that gases obey the ideal gas laws fairly accurately at such densities that the mean intermolecular distances are not extremely large compared with the separation in the solid phase, shows that the attraction must decrease very rapidly with distance. There is an abundance of both experimental and theoretical information on this question. The results for the interaction of two molecules are represented qualitatively in *Figures 4* and *5*.

For a system of two molecules there is an equilibrium separation r_0 at which the force of interaction is zero. At greater distances the force must be attractive, at lesser distances it is repulsive, hence the equilibrium is stable. In a large assembly each molecule must interact with a large number of others. To a first approximation we need only consider the interactions with near neighbours and we will therefore expect that molecules will tend to be arranged such that the mean distance between near neighbours is very roughly equal to r_0. Obviously the detailed interactions of a

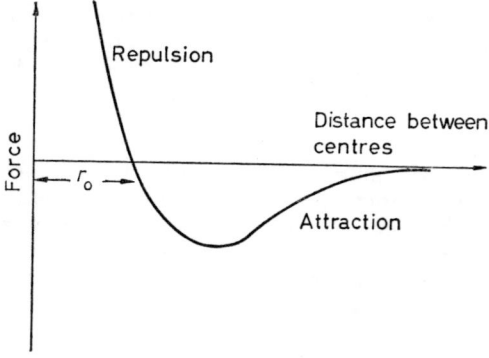

FIGURE 4.

three-dimensional system are very complex, but several of the main features can be understood in simple terms by reference to the behaviour of a pair of molecules.

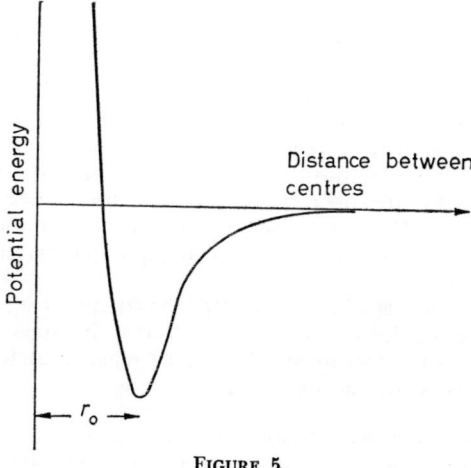

FIGURE 5.

At certain points in a solid or liquid the molecules will be subject to zero force. On displacement there will be restoring forces which cause oscillation about these points. From *Figure 4* it can be seen that for small displacements Hooke's law applies and the oscillations will be S.P.M., their energy being half kinetic and half potential. For large amplitudes the simple laws break down.

Elasticity

If a condensed substance is compressed the molecules are driven closer togther and the mean distance between near neighbours becomes less than r_0, hence the mutual forces are repulsive. If it is stretched the mean distance becomes greater than r_0 so the forces are attractive. This gives to the material elasticity with Hooke's law applicable for small strains.

Elementary texts on mechanics attempt to describe elasticity in terms of molecules with attractive forces only. This leads to the same absurdities as were found in the accounts of surface tension discussed in Chapter 1. A typical account states 'It is the powerful attraction between the molecules which gives a solid its elasticity'. It should be obvious that if there were only attraction matter would collapse and that repulsion is equally necessary for a material to be elastic.

Another typical example of the traditional treatment of intermolecular forces is as follows. 'If two steel blocks have their surfaces ground very flat and placed in contact they will cling together strongly because the molecules in one block are attracting those in the other.' Normally of course if two blocks are in contact there is no resultant force between them. If we attempt to pull them apart there is an attractive force, and if we press them together there is a repulsive force.

Similar considerations apply to the frequent use of the terms cohesive force and adhesive force in such contexts that the mean forces must be zero. In most cases it appears that what is really meant is *energy*.

Surface Energy and Latent Heat

Let us suppose that each molecule in the interior of a

liquid or solid has N nearest neighbours. Since the potential energy changes very rapidly with distance we can for a first approximation ignore interactions with more remote particles. Hence the molecule is bound to the condensed phase by an energy approximately equal to NU where U is the energy required to separate it from one neighbour. To remove a molecule it is necessary to supply this amount of energy, which generally accounts for the greater part of the latent heat of evaporation. (A few parts per cent of the latent heat of sublimation of a solid are required to overcome the external pressure, whilst for a liquid the fraction is generally rather higher, being typically ten per cent at the boiling point).

A molecule in the surface has only M near neighbours where M is less than N. This gives a 'free surface energy' equal to $(N-M)U$ per molecule in the surface. This energy will clearly be roughly proportional to the latent heat. The weaker binding of the surface molecules does not usually affect evaporation since the removal of a surface molecule will bring another molecule from the interior to the surface, so the work done is NU. The surface energy is however important when the shape of a body changes, altering the proportion of molecules which are on the surface. Clearly work must be done to stretch the surface, hence there is surface tension.

Thermal Expansion and Electrical Resistance

It is commonly taught that the thermal expansion of condensed materials results simply from the increase with temperature of the amplitude of the molecular oscillations. It is assumed that an oscillating molecule must occupy more space than one at rest, but in the case of a simple periodic oscillation the motion is perfectly symmetrical and the mean position of the particle is the mid-point, hence there can be no expansion in this case. At low temperatures the amplitudes of oscillations are small and the motion is S.P.M., hence expansion coefficients tend to zero. At higher temperatures the molecules oscillate into regions where Hooke's law is no longer applicable. From *Figures 4* and *5* it is apparent that such 'anharmonic' oscillations cause an increase in the separation of near neighbours, giving thermal expansion.

As would be expected from this argument, expansion coefficients of solids are not constants but generally increase with rising temperature.

Another phenomenon which depends upon the departure of the oscillations from S.P.M. is the temperature dependence of resistivity of metals. It is sometimes taught that an increased amplitude of oscillation of an ion in some way makes it occupy more space and increases the chance of an electron hitting it. However, an oscillating ion is just as likely to move out of the way of an electron as into its path. The oscillation will only affect the probability of collision if it is not S.P.M. In this case the mean distance of the ion from its near neighbours is increased and there is a local distortion of the regular crystal structure. It is such distortions that scatter electrons as they move through a material.

The Purpose of the Kinetic Theory of Gases

Elementary physics courses use kinetic theory firstly in qualitative accounts of phenomena and secondly in deriving the formula for the pressure P of a gas

$$P = \tfrac{1}{3}\rho c^2$$

where c is the root-mean-square velocity of the molecules of an ideal gas of density ρ. In view of the properties which are usually attributed to the molecules it is necessary to question the purpose in teaching the theory.

Perhaps it is the intention to give the student some idea of the properties of molecules and to show how the behaviour of matter in bulk can be explained in terms of the interactions of its constituent particles. If this is the object the conventional approach is completely unsuitable since, as will be shown, the properties that are attributed to the molecules are mostly incorrect. An alternative purpose could be to conduct an intellectual exercise showing how a particular mechanical model can be used to interpret a few phenomena, without regard to the facts that exactly the same results can be obtained with completely different assumptions and that the properties assumed for this model are inconsistent with clearly demonstrable properties of molecules. Such an object might be legitimate, but if it is pursued it is obviously essential to make it quite clear to the student that this is in fact the purpose of the analysis.

Properties of Molecules

It appears that most teachers of this subject endeavour to fulfil the first purpose stated above, and suppose that the properties they attribute to molecules are in fact necessary to account for the properties of matter. Thus it may be stated that 'These results have been obtained on the assumption that molecules are elastic spheres. If this assumption is not true the predictions of the theory will not agree with experiment'. We must examine the assumptions of the theory from this point of view.

Fairly commonly it is assumed that molecules are spherical. The idea is that a spherical molecule will bounce off the surface of a container with the angle of reflection equal to the angle of incidence. Such an analysis of course assumes that the surface is flat on a molecular scale, that is that solids are not made of molecules, although gases are. The pointlessness of this assumption is shown by the fact that, on chemical and physical evidence, the hydrogen molecule is very far from spherical but hydrogen obeys the elementary gas laws more exactly than nearly all other gases. Also much research on gases at low pressures has shown that molecules of all shapes are reflected by surfaces in random directions. Obviously any result relating to a gas in equilibrium in a container can equally well be given by the two quite different assumptions (*a*) that molecules undergo mirror-type reflections, or (*b*) that molecules are randomly scattered. The latter assumption is justified by the evidence, which shows that the shape of the molecules is irrelevant to this question.

A more common assumption is that molecules are perfectly elastic, that is kinetic energy is conserved in all collisions. Again, this assumption is not necessary to obtain any result for gases in equilibrium, since all that is necessary is to assume that the effects of inelastic and superelastic collisions must balance out. There is in fact clear evidence that collisions are not generally elastic in the cases of diatomic and more complex molecules. The values of the specific heats show that such molecules have amounts of rotational and vibrational energy which depend on temperature. Inelastic and superelastic collisions must occur to transfer such energy to and from the molecules on changing the temperature. That such collisions are very common is shown by the fact that the specific heats have the same values for rapid changes of temperature, giving constant speeds of propagation for ultrasonic waves up to very high frequencies.

In the derivation of the simple gas laws by kinetic theory there is one essential physical assumption which must be made, namely that the range of intermolecular forces is small compared with the mean distance between molecules. This is usually stated in one of the forms (*a*) 'intermolecular forces are neglected' or (*b*) 'the range of intermolecular forces is zero'. Now if there were no intermolecular forces the gas could not possibly exert a pressure. Similarly, if the range of the repulsive forces were zero the molecules would penetrate the walls of any container. Actually in very many accounts of kinetic theory the writers only mention attractive forces, and it is obviously believed that a gas exerts a pressure on a wall without there being any repulsive intermolecular force.

The Effects of the Walls

In kinetic theory the effects of the walls upon the gas are assumed to be independent of the particular properties of the molecules of the material of the wall. This is physically necessary, since the gas laws are identical in containers of different materials. This simple point causes considerable difficulties. There are two very different explanations given in textbooks. According to one account the walls are coated with a layer of gas molecules which by some mysterious process suppress the attractive forces between the wall and the molecules in the gas. On this assumption one would expect the repulsive forces to be suppressed similarly, in which case the gas would exert no pressure and would leak out from the container.

The other explanation is quite different. It points out that as a molecule approaches the wall it will be attracted towards it, thereby gaining momentum. On collision with the wall this momentum is given back, producing zero resultant effect. It should be obvious that the same argument applies to the collision of two molecules in the gas, but in this case the existence of attractive forces is correctly assumed to affect the equation of state. The fallacy of the argument should be apparent on considering the potential energy of the molecules. If the range of the attractive forces of the walls were significant compared with the dimensions of the vessel the gas equations would certainly be affected by the nature of the wall material. To expand the gas, work would have to be done to reduce the number of molecules near the walls, hence the pressure would be influenced by the range and strength of these forces.

The reason why we do not have to consider the properties of the walls is that the range of the intermolecular forces is in every case very short compared with the dimensions of the container.

Intermolecular Forces in Different Phases

It is taught almost universally that intermolecular forces are strongest in solids, much weaker in liquids, and weaker still in gases. The truth is the reverse of this. For a given substance the strength of the intermolecular forces is, as we have seen, a function of distance. In a solid the molecules oscillate with small amplitudes about points at which the forces are zero. Rarely does a molecule become displaced sufficiently for a very large force to act on it. At higher temperatures the substance becomes liquid and the molecules generally have larger amplitudes of oscillation involving larger restoring forces. In a gas at low pressures the molecules do not interact significantly for more than a small fraction of the time. When they collide, however, the forces involved are much stronger than those normally arising in condensed phases.

It is probable that this point is misunderstood by confusion between force and energy. The mean potential energy of the interaction between molecules has a high negative value for solids, it is a few per cent less negative for liquids and is relatively nearly zero for gases at low pressures.

The Equation of Van der Waals

Kinetic theory is consistent with the equation of state $PV = RT$ upon the assumption that the ranges of the intermolecular forces are small compared with the mean distances between the molecules. For gases of moderate or high density, and for liquids, this assumption is not valid and more exact equations of state, such as that of Van der Waals, are needed.

The equation of Van der Waals is derived in many textbooks. There are two points to consider, namely the effects of the repulsive intermolecular forces at very short distances and of the attractive forces at slightly larger distances. The effect of the repulsive forces can be considered to a first approximation as a reduction of the volume within which molecules are free to move. This is usually expressed by saying that the molecules themselves occupy a certain volume. Now in elementary courses intermolecular forces are nearly always said to be attractive, and the volume of a molecule is not associated in any way with a repulsive force. Quite clearly it is thought that there is some process by which a molecule can prevent another from entering the same region of space as itself without exerting any force on it.

The attractive forces are usually taken into account in the following way. The gas is imagined to be divided into layers which attract each other, whilst the outer layers are supposed not to be attracted by the walls of the container (p. 24). In consequence it is concluded that the 'true pressure' inside the gas is greater than the 'observed pressure' at the walls. No meaning is normally attributed to the 'true pressure', nor does it seem that this phrase as used has any meaning. The pressure of a gas *in equilibrium*, represented by the symbol P in the equation of state, is the normal force per unit area exerted by the gas on any surface. This is the observed pressure, and is the only quantity which can be said to be the *actual* or *true* pressure. If, as is supposed, the layer of gas next to the surface of a wall were to exert a higher pressure on the gas layer next to itself than that it exerts on the wall, it would be accelerated outwards. This is of course precisely what occurs in a gas which is beginning to expand, but in that case the system is not in equilibrium and P has no definite meaning.

The equation of Van der Waals is

$$\left(P + \frac{a}{V^2}\right)(V-b) = RT$$

where the term a/V^2 accounts for the attractive intermolecular forces and b accounts for the repulsive ones. The quantities a, b and R are constants for a given mass of a particular fluid. The traditional analysis of the effects of the attractive forces give this equation with the power of V in the first bracket equal to minus two, independently of the law of dependence of force upon distance, which is not involved in this treatment. Sometimes, however, it is supposed that the value of the power of V is a consequence of the attractive forces obeying the inverse-square law, and it is even thought by some people that these forces are gravitational. Now it can be shown thermodynamically that the power of V in the equation of state is in fact dependent upon the law of force. The analysis is worthy of consideration here, although it is beyond the standard of sixth-form physics.

The actual law of force is represented by the potential energy function shown as a full curve in *Figure 6*. As an approximation this is replaced by the mathematically simpler form shown by the dashed curve. The repulsive forces are supposed to tend to infinity whilst the distance over which they operate tends to zero giving the vertical broken line in the figure. The potential energy is then represented by a simple power function of the distance of the form r^{-m}. Clearly the value of m will be less than the numerical value of the power in the equation for the full curve. The broken curve gives approximately the average effect of the resultant of the attractive and repulsive forces in the region of negative potential energy.

According to the laws of thermodynamics an infinitesimal

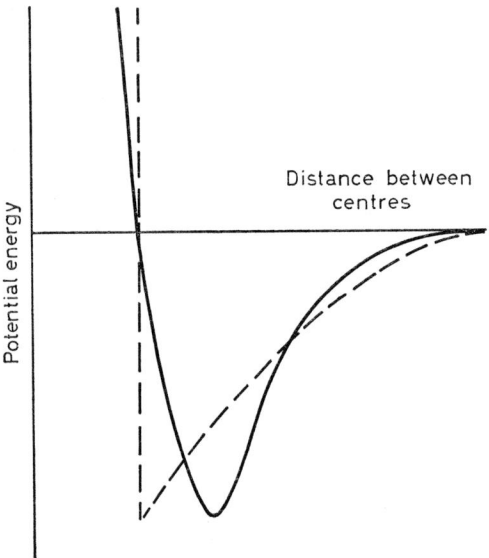

Figure 6.

volume change dV causes a change dU of the internal energy given by

$$dU = TdS - PdV$$

where dS is the change of entropy. The equation can be written in partial form with T constant

$$\left(\frac{\partial U}{\partial V}\right)_T = T\left(\frac{\partial S}{\partial V}\right)_T - P$$

The function of entropy can be changed in the usual way into a function of temperature using one of Maxwell's relations, hence

$$\left(\frac{\partial U}{\partial V}\right)_T = T\left(\frac{\partial P}{\partial T}\right)_V - P$$

Differentiating the equation of Van der Waals this gives

$$\left(\frac{\partial U}{\partial V}\right)_T = \frac{a}{V^2}$$

As the volume V must be proportional to the cube of the intermolecular distance r this equation gives

$$\left(\frac{\partial U}{\partial r}\right)_T \propto \frac{1}{r^4}$$

Hence the index m of the equation of the broken curve is equal to 3. There are theoretical reasons for believing that the full curve is given by the law of the inverse sixth power.

Although this analysis is necessarily approximate it is important in showing that the power of V in the first bracket of the equation of Van der Waals is dependent on the effective law of force. Hence the traditional treatments which omit any consideration of the form of this law are invalid. The thermodynamic treatment of this problem was first published in 1886.

Absolute Zero

It is taught almost universally in elementary courses that molecular motion ceases at the absolute zero. That this is not the case should be apparent from a simple application of the indeterminancy relations. If a molecule is at rest the indeterminancy in its momentum is zero, hence by Heisenberg's principle the indeterminancy in its position is infinite. Hence there must be 'zero-point energy' in all bodies in which the molecules are constrained to a finite volume, that is in all real bodies. A much more exact treatment is possible for solids using the Debye theory of specific heats, from which it is easily shown that the energy of atomic vibrations in a gram-atom of a solid at absolute zero is $(9/8)R\theta$ where R is the universal gas constant and θ is the Debye temperature. For a gram-atom of iron, with θ equal to about 450°K, this gives a zero-point energy for the oscillation of the ions of over four thousand joules. The total zero-point energy is actually far greater even than this since in metals the value for the conduction electrons is greater than that for the ions.

The idea that molecular energy vanishes at absolute zero probably arises from a misunderstanding of the properties of ideal gases. From elementary kinetic theory it can be shown that if the range of the intermolecular forces is small compared with the mean intermolecular distance the translational kinetic energy of a gram-molecule of a gas is $(3/2)RT$ where T is the Kelvin temperature. For such a case clearly the kinetic energy would be zero at absolute zero. However, the conditions of relatively very small ranged forces implies vanishingly small pressure, hence extremely large volume. The ideal gas laws only apply exactly in the limit when $P \to 0$ and $V \to \infty$. Now for a gas of infinite volume the position of a molecule can be infinitely indeterminate. Hence the momentum can be exactly determinate and it is possible for the kinetic energy to be zero. Obviously this is not the case for any real substance.

The theory of specific heats shows that the relationship between molecular kinetic energy and temperature is complicated and is different for different materials. It is true to say that temperature is 'a measure of' the molecular kinetic energy, but it is wrong to assume any simple correspondence. In general this energy is not proportional to the Kelvin temperature, nor is it zero at $0°K$. Further, the molecular kinetic energy does not generally remain unchanged in an isothermal change of phase.

The idea of absolute zero is very difficult and it cannot be explained in simple terms. Nothing but harm can result from our teaching a completely false idea of this subject.

Chapter 5

TEMPERATURE

Scales of Temperature

The concept of temperature arises from sensory experience, but this gives only qualitative ideas. To define temperature quantitatively it is necessary to refer to some continuously variable property of a material. Each physical property of each material used defines a distinct scale of temperature, hence there is effectively an unlimited number of scales.

The unit of a scale is defined by giving arbitrary values to certain fixed points. The most familiar of these are the ice point and the steam point. The ice point 0°C is the temperature of ice in equilibrium with water at standard atmospheric pressure, the material being pure except for dissolved air with which it should be saturated. The steam point 100°C is the temperature of steam in equilibrium with pure water at standard atmospheric pressure.

To define a temperature t on any particular Celsius scale some property of a material is measured at this temperature and at the ice and steam points. Let the measured values be X_t, X_0 and X_{100} respectively, then by definition

$$t = \left\{\frac{X_t - X_0}{X_{100} - X_0}\right\} \times 100°C \qquad \ldots \ldots (5.1)$$

It is essential to note that a particular temperature will have different numerical values on different scales. Only at the fixed points can we be sure that any two scales will agree. The 'size' of a Celsius degree will vary according to the scale used, the only common feature being that the *average* size of the degrees between 0°C and 100°C must be the same on all scales. For example, consider two scales based on the electrical resistances of coils of two different materials. A certain fixed point might be 40°C on one scale and 41°C on the other. Clearly the degrees of the two scales are unequal over the range between this temperature and the ice-point. At a very high or very low temperature the two types of degree might even have different orders of magnitude.

Most temperature scales only have meaning over a limited range of values. For example, there is no such thing as temperature on a mercury-in-glass scale below the freezing point of mercury. Clearly it is meaningless to extrapolate any law expressed in terms of such a scale beyond the limits within which it is logically definable.

The Ideal Gas Scale

Until about 1850 most measurements of temperature were made using mercury-in-glass thermometers, and physical laws were expressed on a mercury scale. As more precise measurements became possible it was found that different mercury thermometers had slightly different scales and it became necessary to seek some more reliable standard. Gas thermometers were found to be more stable than mercury ones, and there was close agreement between different gas scales over a wide range of temperatures. This agreement was found to be particularly close when working with low gas pressures. Lord Kelvin proposed the use of an ideal gas scale based upon the value of the product PV of a gas, extrapolated to zero P. He showed that all the extrapolated gas scales derived from measurements on real gases must be identical, as is explained in the next section. This then provides a fundamental scale which is the standard usually employed for theoretical purposes. This scale has an absolute zero at a finite number of Celsius degrees below the ice point, and the absolute zero is now taken as one of the fixed points on the Kelvin scale. The size of the degree is fixed by defining the temperature of the triple point of pure air-free water to be $273 \cdot 16°K$. (This is very nearly $0 \cdot 01°C$.) The International Practical Temperature Scale is a Celsius scale so defined as to agree as closely as possible with the absolute ideal gas scale.

Thermodynamic Scales

Consider a series of ideal reversible heat engines so arranged that the waste heat from one is fed reversibly to the next, as represented in *Figure 7*. Q_N is the heat supplied in unit time at temperature T_N to the Nth engine which does work $W_N = \varepsilon_N Q_N$. By definition ε_N is the efficiency. The waste

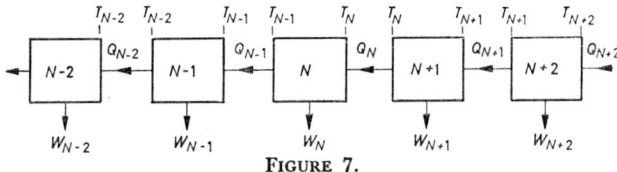

FIGURE 7.

heat Q_{N-1} is discharged into the engine number $(N-1)$ at temperature T_{N-1}.

We have
$$Q_{N-1} = Q_N - W_N = Q_N (1 - \varepsilon_N)$$

By Carnot's theorem the efficiencies of such ideal engines are functions of temperature only, hence Kelvin proposed that temperature scales could be defined in terms of these efficiencies. Such scales would be fundamentally independent of the properties of any particular substance, although measurements upon real materials would be needed for practical application. Many such thermodynamic scales could be defined, but there are two obvious simple possibilities.

(1) The intervals between intake and output temperatures for the engines are defined to be equal if the efficiencies are equal; or

(2) The temperature intervals are defined to be equal if each engine does the same amount of work.

In either case the size of the degree can be determined by calibrating at the ice and steam points, provided some property of matter can be found which has a determinable relationship to temperature on these scales.

These two scales are equally valid in principle, but differ greatly from each other except at temperatures very near to the fixed points. In particular the series of engines required to define the scale *(1)* must extend infinitely to lower and lower temperatures. As the engines have equal efficiency there is always waste heat to be discharged at the lower temperature. Hence the absolute zero on this scale is at minus infinity degrees Celsius. The scale (2), however, demands a finite number of engines below the Nth, since each engine does an equal finite amount of work W_N and the heat input is finite. Hence the absolute zero on this scale is at a finite number of degrees Celsius below the ice point. Thus we see that the existence of an absolute zero at a finite negative temperature on the Celsius graduation is not a necessary property of temperature scales but purely a characteristic of certain types of scale.

Consider the last few heat engines of the series defining the scale (2) on going to the lowest temperatures, as shown in *Figure 8*. The scale is meaningless below that temperature

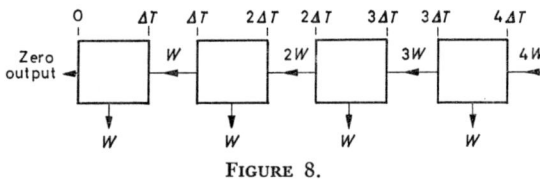

FIGURE 8.

at which zero waste heat is discharged, which is clearly absolute zero. The heat input to the last engine, working between the absolute temperatures 0 and ΔT must be equal to the external work done W, hence the efficiency is unity. The next engine works between $2\Delta T$ and ΔT, the heat intake must be $2W$ and the efficiency is a half. The third engine has efficiency one third, and so on. By inspection we can deduce the law for the efficiency ε

$$\varepsilon = 1 - \frac{T_2}{T_1}$$

where T_1 is the input temperature and T_2 the output temperature on this scale.

Now it is easily shown that the efficiency of a Carnot type engine using an ideal gas as working substance is given by

$$\varepsilon = 1 - \frac{\theta_2}{\theta_1}$$

where θ_1 and θ_2 are the input and output temperatures measured on the absolute ideal gas scale. Hence we can identify any absolute ideal gas temperature θ with the Kelvin thermodynamic temperature T on scale (2) by using the same fixed points.

The ideal gas scale can be realized very precisely by experiments on real gases at low pressures, and for rough work the scale is in close agreement with the familiar mercury thermometers over the range of utility of the latter instruments. The thermodynamic scale (2) has thus been adopted generally as the standard theoretical temperature scale. It should be understood, however, that other theoretical scales are possible, and for some purposes might be preferable.

From this analysis it should be apparent that the procedure of converting Celsius scale into an absolute one by adding 273 degrees is only meaningful if the scale is based on the ideal gas, and the added degrees are also on this scale.

Traditional Treatment of Temperature Scales

In many elementary courses it is not indicated which temperature scale is used. It is normal to attribute to temperature scales in general, properties which are only applicable to particular scales. It is taught that the interval between the ice and steam points is 'divided into one hundred equal parts'. Such a statement in the abstract is meaningless, since temperature increments which are equal on one scale are usually unequal on others.

It is common to explain the choice of mercury as a thermometric substance on the grounds that 'the expansion of mercury is uniform'. Uniform with respect to what? By definition it must be uniform with respect to a mercury scale, but the expansion of aluminium must be uniform with respect to an aluminium length scale, and so on. The expansion of mercury is not uniform when expressed in terms of a temperature scale defined by the radiation from a black body or by the E.M.F. of a copper-iron thermocouple. Every property of matter which varies continuously with temperature can be used to define a temperature scale, and each property is 'uniform' or 'linear' with respect to the scale it defines. Equally, each property is non-uniform with respect to other scales. Presumably the teacher has in mind some other temperature scale in terms of which the expansion of mercury is linear. If, as sometimes seems to be the case, this is a gas thermometer scale it would be illogical to use the mercury thermometer to study the properties of the gas, as is often done.

Further confusion arises from the relationship between internal energy and temperature. Temperature is said to be a measure of the amount of 'heat' contained by a given mass of a particular material, and this sometimes leads to the idea that equal increments of heat imply equal steps in temperature. This idea is explicitly taught when specific heat is defined to be a constant. Hence many students have in mind a 'calorimetric' temperature scale, and when they are taught that a material expands uniformly they imagine that the expansion is referred to this scale. It can be very disturbing to such students to learn that specific heats are in fact functions of temperature.

The extremely difficult concept of absolute temperature is often taught before the ideal gas scale is discussed. As has been shown in the previous section the usual ideas on this subject are not applicable to temperature scales in general, hence this sequence in teaching is most unsuitable. It is sometimes supposed that the idea of absolute temperature must be taught at an early stage to permit the correction of the measured volumes of gases to S.T.P. This is quite unnecessary. All that is required is that the expansion coefficient α of the gas is measured in terms of any convenient thermometer. Then the observed volume V at temperature t is related to the volume at the ice point by

$$V = V_0 (1 + \alpha t)$$

Lengths and volumes of solid and liquid materials often have to be corrected to a standard temperature, and this is done without any reference to absolute temperature by teachers who find this concept necessary when working with gases.

The Expansion of Gases

In nearly every elementary physics course the student

either performs or has described to him experiments in which the volume of air at constant pressure and the pressure at constant volume are determined at various temperatures measured with a mercury thermometer. Graphs are plotted of pressure against temperature and volume against temperature, and the graphs are extrapolated to zero pressure and zero volume respectively. The points at which the lines cut the temperature axis are said to represent absolute zero.

It will be sufficient to discuss the case of the variation of volume, the pressure being held constant. The significance of the traditional procedure is very obscure. Firstly, temperature on the scale used becomes meaningless below $-39°C$ on the mercury scale at which point mercury freezes, so points on the line beyond this temperature have no definable physical meaning. Further, it is known that air condenses at low temperatures. What meaning is to be given to the 'volume which air would occupy if it were a gas—but it is not'? It is said that the point where the volume becomes zero must be absolute zero, since at lower temperatures the volume would be negative, which is inconceivable. However, we know that the material is no longer a gas at very low temperatures so what does this signify? At finite pressures any real substance must condense ultimately to a *finite* volume and hence there is no justification in going to zero volume.

Since a mercury scale is used it is odd that it is assumed that an absolute zero on this scale is defined by the properties of air. If one believes in mercury temperature below the freezing point it is logical to extrapolate to zero volume of liquid mercury, at $-5,500°C$, and call this point absolute zero. Nobody does this, but it is no more absurd in principle than what is actually done.

A clearer idea of the dangers of such extrapolation may be obtained by considering the results which would be found by using a thermometer which can in principle operate down to the lowest temperatures, and which will agree closely with the mercury scale at ordinary temperatures. A suitable scale, such as has actually been used for other purposes, is one defined in terms of the length of a metal rod. The value of the product PV extrapolated to zero gas pressure would be determined at various temperatures t measured on the metal length scale. The curve for PV plotted against t would be as shown in *Figure 9*. The curvature of the graph is caused

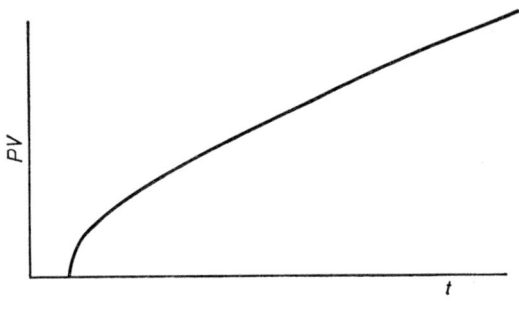

FIGURE 9.

by the temperature variation of the expansion coefficient of any solid when referred to the ideal gas scale. Absolute zero on the metal length scale is obviously given by the actual intercept of the curve where PV is zero for the ideal gas, not by the intercept of the tangent drawn at any higher temperature. If we require the value of the absolute zero on the ideal gas Celsius scale only the readings of PV at the ice and steam points are relevant.

The Law of Charles

The thermal expansion of gases at constant pressure and the pressure variation at constant volume were investigated by several scientists during the latter part of the 18th century and the first few years of the 19th. The results are expressed in various forms and associated with the names of several workers. We shall consider only the variation of volume and describe the result as the law of Charles.

The pioneer workers studied the relationships between pressure, volume and temperature for real gases at finite pressures, using mercury thermometers. Thus the law of Charles is an approximate relationship between the expansion coefficient of mercury and that of a gas. This simple and essential point is seldom if ever stated in elementary courses, in which the law is given as a relationship between the volume of a gas and temperature in the abstract, without reference to any particular scale. Obviously this cannot be meaningful, since different temperature scales can differ so radically from each other. For example, if the first thermodynamic scale defined on p. 33 were used the volume of a gas at constant pressure would be approximately exponentially dependent on temperature. It is also usually uncertain whether the law is in fact an experimental one or is a defined property of an ideal gas. This difficulty has become more serious in recent years as, incredible as it may seem, some teachers and authors now discuss this subject without ever distinguishing between real and ideal gases.

When the laws of Boyle and Charles have been taught it is usual to 'derive' the ideal gas equation $PV = RT$ by combining them, and this derivation is sometimes explicitly demanded by examination boards. The whole procedure is illogical. The law of Charles depends partly upon the properties of mercury, but the ideal gas does not depend in any way upon this substance. The ideal gas equation is by definition an exact relationship for an imaginary substance at all temperatures. The law of Charles is an approximate relationship for real substances over a limited range of conditions. It is impossible to derive a definition, and it is impossible to prove an exact relationship by assuming an approximate one.

The author has asked one examination board what answer was expected to a question on this subject. The examiners stated that they understood the law as 'V is proportional to the absolute temperature T if P is constant'. Now this expression involves the ideal gas temperature T which is defined by the ideal gas equation. Thus the student is expected to assume tacitly that $PV = RT$, since otherwise T has no meaning. From this it is deduced that V is proportional to T if P is constant, and from this the student has to deduce that $PV = RT$. Such an argument in a circle would be unacceptable in second-form mathematics, but is commonplace in A-level physics.

Conclusion

The ideas of temperature, and particularly absolute temperature, are extremely difficult even when taught in a logically consistent way. The subject is made almost incomprehensible by mistaken treatment, particularly the use of several essentially meaningless procedures and the failure to distinguish between theory and experiment and between real and ideal cases. To avoid these difficulties it is proposed that the subject could be taught according to the following scheme.

In elementary work (up to O-level) such difficult ideas as gas laws and absolute temperature should be omitted, and no precise laws explicitly involving temperature should be taught. Temperature should be defined in terms of equation (5.1) and it should be explained that different properties of different materials define different scales. The mercury thermometer would be used for most purposes, and it would be explained that all results obtained with it depend upon certain properties of mercury. Specific heats must be known to be functions of temperature.

At A-level the law of Charles should be taught clearly as an experimental result, without any extrapolations or other invalid deductions. The idea of a gas thermometer can be explained, and it should be taught that the different gas scales are found to agree closely when working at low gas pressures. Use should be made of the simple equation for low pressures

$$PV = A + BP$$

where A and B are found to be constants for a given gas at constant temperature. From this one can introduce the idea of an ideal gas as the limiting case of a real gas as P tends to zero. Hence the ideal gas absolute temperature T can be defined by the equation $PV = RT$. This equation can also be used as an approximate equation of state for real gas, but it is *essential* to distinguish clearly between the real and

CONCLUSION

ideal cases. It should be stated that there is a theoretical proof (too advanced for this level) that all ideal gases must define the same scale. Once the idea of the ideal gas scale is established the subject can be developed without any logical difficulties, all precise laws being referred explicitly to this scale.

Chapter 6

THERMODYNAMICS

Thermodynamic Equilibrium

In the study of the thermal properties of matter it is necessary to consider bodies in thermodynamic equilibrium, that is to say in a state which is completely uniform and which does not change with time. If there is any significant gradient of temperature or pressure, or of electric or magnetic field intensity, then the substance is not in equilibrium even though the state may be steady. Unfortunately the terminology is confused by the frequent use of the term 'dynamic equilibrium', sometimes to mean thermodynamic equilibrium as defined above, and sometimes to mean a steady but non-uniform state.

It may be necessary to explain that when bulk matter is said to be uniform or steady this description only applies to the macroscopic properties. The individual molecules are of course in a state of constant agitation.

In very many applications of thermodynamics and in nearly all basic theory such influences as electric and magnetic fields can be assumed to be negligible and the only constraint considered is a uniform pressure P. For solids, other mechanical constraints such as tension are possible, but these are usually negligible. Solids generally cannot be obtained in states of perfect equilibrium as a result of the very slow relaxation of various defects and non-uniformities. Careful annealing may bring a solid into a condition very close to equilibrium, in which case thermodynamic results which only apply strictly to fluids may be used with good precision.

Consider a given mass of a pure substance in thermodynamic equilibrium. The Kelvin temperature T, the specific volume V and the pressure P can be measured. The specific volume is the volume of unit mass and is therefore the reciprocal of the density. The pressure is the normal force per unit area on a surface in contact with the substance *in*

THERMODYNAMIC EQUILIBRIUM

equilibrium. Provided there is only one phase it is found that if the values of P and T are fixed then V is uniquely determined. The state of the substance is thus determined by the values of pressure and temperature, and the mathematical relationship between P, V and T for a substance in equilibrium is called the equation of state. It is very important to appreciate that an equation of state can normally only be applied to matter in equilibrium, since in other circumstances no simple meaning can usually be given to P, V and T.

When the values of P and T are fixed for a substance in equilibrium then the values of various other quantities as well as V are determined. One of these is the internal energy U, that is the aggregate of the energies of the individual molecules resulting from their random motions and interactions and from their internal structure. For many purposes the latter term can be ignored since the structure of a molecule usually only changes at high temperatures. The zero of internal energy is arbitrary. The value of U and similar 'functions of state' is usually referred to unit mass of material.

Internal energy is sometimes called 'heat' in elementary courses, although in more advanced work this word is only used to mean energy transferred to or from a body by a gradient of temperature. Thus heat can be given to a body by conduction, free or forced convection or radiation from bodies at higher temperatures, but the word heat is not properly used to describe energy supplied by mechanical processes such as compression. According to this terminology it is not correct to speak of the amount of heat in a body.

The teaching of this subject is often confused by failure to distinguish between heat and internal energy. Further confusion arises from the idea, which is very often taught, that heat is a form of motion. Usually it appears that heat is in this context used to mean internal energy. Now in general U is a mixture of kinetic and potential energies, typically in roughly equal amounts, and it is difficult to see that such a function is 'motion' in any sense. The flow of heat in the more strictly correct sense of the word must involve motion, but again it is not clear that heat *is* motion.

Reversible and Irreversible Changes

Consider a system which is initially in a state of equilibrium and is then changed to a different state, the final state also being in equilibrium. The change is said to be *reversible* if at every stage in the process the system is in equilibrium. Since an equilibrium state is steady by definition such a process can in principle only take place infinitely slowly, hence the term 'quasi-static' is sometimes used. A change at a finite rate requires the acceleration of certain parts or gradients of temperature or pressure which will destroy equilibrium. An effectively reversible process can occur at a finite rate if there is no resistance to flow of matter or heat, that is friction is zero and thermal conductivity is infinite. In this case an infinitesimal change of conditions will reverse the direction of change *exactly in every detail*. This is the implication of the term *reversible*.

Clearly few real processes can be exactly reversible in the thermodynamic sense. In some circumstances it is possible to cause changes which are very nearly reversible, an important example being the alternating volume change in a sound wave. In this case the magnitude of the change is extremely small, so that although the process is very rapid departures from equilibrium are infinitesimal.

Real changes are normally irreversible. If a process occurs spontaneously the initial state cannot be one of equilibrium, hence the change cannot be reversible. Even if the initial and final states are both under equilibrium conditions, in some intermediate stage there will usually be a significant departure from perfect equilibrium, especially if the change is rapid. It is of course often possible to restore the material to its original state, but this cannot be done by an exact step-by-step reversal of the process of change. It is a very serious error to suppose that a process is reversible just because the material concerned is restored to its initial condition.

Expansion Processes

Consider the *reversible* expansion of unit mass of a sub-

EXPANSION PROCESSES

stance. Since the material is in equilibrium the normal force per unit area exerted upon the surroundings is the pressure P of the equation of state. If a surface of area A is driven back through a distance dx the external work done dW is equal to $PAdx$ which is equal to PdV where dV is the change of the specific volume.

Thus

$$dW = PdV \qquad \ldots \ldots (6.1)$$

and for a finite expansion

$$\Delta W = \int_{V_1}^{V_2} PdV \qquad \ldots \ldots (6.2)$$

The value of the integral depends upon the particular reversible path taken from V_1 to V_2, so W is not a function of state.

It is essential to appreciate that equations (6.1) and (6.2) only apply to reversible processes. In a real expansion there will be accelerations and also dissipative resistance to motion, the latter arising from viscosity and turbulence in a fluid and from elastic hysteresis in a solid. Since the substance is not normally in an equilibrium state during an irreversible expansion it is not easy to give any definite meaning to the pressure, but it should be clear that the normal force per unit area on a surface being driven back by the expanding material will be less than that which would be exerted in a reversible expansion, other things being equal. For an infinitesimal irreversible expansion, the initial and final states being in conditions of equilibrium at pressure P, we will have

$$dW < PdV$$

The work done in an expansion can thus have any value from PdV in an ideal reversible case down to zero in the most unfavourable case. This extremely important fact is rarely taught in elementary physics and equations (6.1) and (6.2) are commonly supposed to apply to every expansion process, although actually they really represent only an ideal limiting case.

It will be noticed that the arguments given above apply equally to gases and condensed matter, contrary to much elementary teaching according to which only gases do work on expansion.

Expansion of Thermally Insulated Substances

The amount of work which can be done in any expansion process may be affected by the degree to which heat may enter or leave the expanding material. One of the most important fields of study in elementary thermodynamics is that of volume changes of materials with perfect thermal insulation, such that no heat enters or leaves the specimen during the process. In practice such insulation often arises from the large bulk of material or from the rapidity of the volume change rather than from the use of lagging material.

Processes which occur with perfect insulation are usually called *adiabatic*. Very unfortunately there is no universally accepted terminology, and many workers limit the term adiabatic to those processes which are perfectly insulated and *reversible*, whereas many others use it to mean *any* perfectly insulated process. This is a cause of grave confusion.

The majority of the cases of expansion with thermal insulation which are discussed in elementary courses depend for their insulation on the rapidity of the change. Unless the expansion is very small it cannot be described correctly by the ideal equations, since if it is very rapid it cannot be reversible and if it is not very rapid it is not properly insulated.

Expansion of Gases with Thermal Insulation

The best known result in this subject is that for the reversible expansion of an ideal gas of constant specific heat. In this case it is found that

$$PV^\gamma = \text{Constant} \qquad \ldots \ (6.3)$$

whence $\qquad TV^{\gamma-1} = \text{Constant} \qquad \ldots \ (6.4)$

where $\gamma = C_p/C_v$, the ratio of the principal specific heats. These formulae are the subject of very many problems in G.C.E. and similar examinations, and they are discussed in many textbooks. In nearly every case the limitations upon the applicability of these equations are completely ignored. Of these limitations the most important is the requirement that the volume change must be reversible. In practice there are a very few processes which can be regarded as reversible, particularly the passage of an acoustic wave and the very small expansions in experiments such as that of Clément and Desormes. Real large expansions such as occur in cylinders and turbines may be described very roughly by equations (6.3) and (6.4). Many other cases including the very important Joule and Joule-Kelvin experiments are not described by these equations even as a crude approximation.

The Joule expansion experiment is performed to seek evidence for the dependence of the internal energy of a gas upon the volume. Two lagged vessels are joined by a tube with a tap. One vessel contains gas at high pressure and the other is evacuated. On opening the tap gas flows until, after various oscillations, it settles down to uniform state. The flow is clearly irreversible because of the finite pressure difference giving acceleration, turbulence and non-uniformity of the material. No external work is done and equations (6.3) and (6.4) are completely inapplicable. If the gas is very nearly ideal the final temperature will be the same as the initial value, hence the formulae for this expansion process are

$$PV = \text{Constant} \quad \ldots \quad (6.5)$$

$$T = \text{Constant} \quad \ldots \quad (6.6)$$

According to the usual treatment in elementary courses equations (6.3) and (6.4) apply to any expansion of a gas under 'adiabatic' conditions, by which is meant normally only that the insulation is perfect. When the Joule experiment is described, however, the results (6.5) and (6.6) are given, without any explanation of why equations (6.3) and (6.4) are not used.

An even more important insulated expansion process is that studied by Joule and Kelvin in the porous plug experiment. This is a continuous process in which a fluid (not necessarily a gas) is driven at a steady rate through a small orifice. The theory assumes that there is negligible change in the kinetic energy of bulk flow and that all processes are reversible except in the vicinity of the orifice. Then if unit mass of fluid is forced through with an input pressure P_1 external work equal to P_1V_1 is done on it by the fluid behind driving it towards the constriction, and on emerging through the orifice at pressure P_2 it does external work P_2V_2 on the fluid in front. The difference between the amounts of work is the change of internal energy if the thermal insulation is perfect. The net external work can be positive or negative or zero and the temperature of the fluid may increase or decrease or remain unchanged. As is shown in books on thermodynamics the temperature change depends upon the equation of state for the fluid, an ideal gas having zero effect.

The equations (6.3) and (6.4) are completely inapplicable in this case also since flow through an orifice is essentially irreversible. Again, it is found that this point is not explained or discussed in most elementary courses.

A particularly odd feature of the teaching of elementary thermodynamics is the fact that several textbooks introduce the idea of reversibility *after* the discussion of expansion processes. In such cases the relevance of the idea of reversibility to the earlier topics is not explained.

In examinations many problems are set upon rapid expansion processes. Such questions can only be answered on the assumption that the processes are perfectly reversible, which assumption is generally applicable only as a crude approximation or not at all in the circumstances described in the questions.

The Principal Specific Heats

When a substance is maintained at constant volume any added heat serves only to change the internal energy so the specific heat at constant volume can be written

$$C_V = \left(\frac{\partial U}{\partial T}\right)_V$$

On heating at constant pressure work is done in the consequent volume change. For the definition of specific heats reversible processes must be considered and equation (6.1) applies. Hence the heat required to raise the temperature by dT is

$$Cp\,dT = \left(\frac{\partial U}{\partial T}\right)_V dT + \left(\frac{\partial U}{\partial V}\right)_T dV + P\,dV \quad \ldots \quad (6.7)$$

$$\therefore \quad Cp = C_V + \left\{\left(\frac{\partial U}{\partial V}\right)_T + P\right\}\left(\frac{\partial V}{\partial T}\right)_P \quad \ldots \quad (6.8)$$

The last term in equation (6.7) represents the external work done in expansion whilst the previous term represents the internal work. For an ideal gas the latter quantity is zero and equation (6.8) gives the well-known result

$$Cp = C_V + R$$

This applies exactly to an ideal gas but only approximately to real gases.

For solids and liquids the internal work is usually far larger than the external work. In the traditional teaching of elementary physics it is only usual to discuss the work done in the expansion of gases, and hence students normally suppose that Cp only differs from C_V in this case. Admittedly the difference is in practice most important for gases, but for typical solids Cp is a few per cent greater than C_V, whilst liquids can have larger differences. These differences are important partly because the ratio Cp/C_V is equal to the ratio of the acoustic bulk modulus to the static bulk modulus, but also because all theories of specific heats give C_V, whilst experimental measurements for liquids and solids give Cp.

Entropy

This concept is not discussed in the majority of sixth-form courses, but it is taught for some scholarship examinations, and in various college courses of about this level. As entropy is of great importance in later studies it is most unfortunate that its treatment at lower levels is often inaccurate. This is particularly serious since it is a difficult and abstract concept and cannot be understood unless it is defined precisely, and explained fully.

The definition of entropy has two essential parts.

(1) The entropy of a body in equilibrium is a function

of the state of that body. Thus whatever is done to the body, if it is restored to the initial state its entropy is restored to the initial value.

(2) If a quantity of heat dQ is added *reversibly* to a body at Kelvin temperature T the increase in the entropy S of the body is given by

$$dS = \frac{dQ}{T} \qquad \ldots \ldots (6.9)$$

Note the following points.

(a) Only *differences* of entropy have been defined. An arbitrary zero is usually taken, although there are reasons for considering a pure substance at $0°K$ to have zero absolute entropy. This raises advanced questions which cannot be discussed here.

(b) The entropy of a body as defined above has the dimensions of heat divided by temperature, the same as thermal capacity. The entropy of a substance is usually expressed per unit mass and hence has the dimensions of specific heat. It is unfortunate that the same name and symbol are given to these two different functions. It might be helpful if a different name, such as *specific entropy*, were used for entropy per unit mass. The distinction is important, since the entropy of a system is the sum of the entropies of the parts, but the specific entropy is not the sum of the specific entropies of the parts.

(c) We omit discussion of the entropy of a system which is not in equilibrium.

(d) From the definition, the value of the entropy of a body does not change in a *reversible* adiabatic process.

(e) For a finite heat increment the temperature may change, then

$$\Delta S = \int \frac{dQ}{T}$$

(f) Entropy has been defined by two separate statements. It is essential to show that these are compatible and that both are necessary.

It is very usual even in degree level courses to define entropy by part (2) of our definition only. It is then a standard exercise to 'prove' that entropy is a function of state, by the following argument. A given mass of a substance is taken round a Carnot cycle, that is a succession of *reversible* changes in the sequence: isothermal expansion; adiabatic expansion; isothermal compression; adiabatic compression. During the isothermal expansion at temperature T_1 the entropy increase of the substance is

$$\int dQ/T_1 = Q_1/T_1$$

where Q_1 is the heat added. Similarly in the isothermal compression at temperature T_2 the entropy given out is Q_2/T_2, where Q_2 is the heat output. From the expression for the efficiency of such a heat engine given in Chapter 5 it can be seen that Q_1/T_1 is equal to Q_2/T_2. As there is no change of entropy on the reversible adiabatics the substance has the original value of entropy on completing the cycle. It can be shown that any *reversible* cycle can be represented as the combination of a number of Carnot cycles, hence on completing any reversible cycle the entropy as defined by equation (6.9) will return to its initial value.

The preceding argument is commonly held to prove that entropy is a function of state. The analysis has however only considered reversible processes. What is the change of entropy of a substance when taken round a non-reversible cycle? The entropy changes are undefined since equation (6.9) refers explicitly to reversible processes only, hence it cannot be proved that the change of entropy is zero. The fact that entropy is a function of state is thus not a result which can be derived from equation (6.9) but is an additional property of entropy which is imposed by definition.

It is important to note that this definition and analysis require the thermodynamic temperature, T, no other temperature being suitable.

The analysis given above, whilst it cannot prove that entropy is a function of state, does show that we are perfectly free to require that the function partially defined by equation (6.9) shall be so. One very important consequence of this requirement is that *dS is not equal to dQ/T for non-reversible processes*. Consider a cycle which is similar to the Carnot cycle except that the adiabatic processes are not reversible. By Carnot's theorem this cycle cannot be more efficient than the perfectly reversible one working between the same temperatures, and generally it will be less so. Thus the efficiency $(Q_1 - Q_2)/Q_1$ is less than $(T_1 - T_2)/T_1$ hence the entropy given out by the working substance in one cycle, Q_2/T_2, is greater than the entropy taken in, Q_1/T_1. As the working substance has finally the same entropy as at the beginning some entropy has been created by the engine. This creation of entropy must occur in the irreversible processes.

When the concept of entropy is taught in scholarship and similar level courses a number of these essential ideas are commonly omitted. Thus it is usual to give some such definition as this 'When heat dQ is added to a body at temperature T the increase of entropy dS is given by $dS = dQ/T$.' Often the *essential* requirement that the process must be reversible is omitted. Now the difference between the changes of entropy in reversible and in non-reversible processes is not just a refinement, but embodies one of the most important and valuable properties of this function. A quantity defined such that its change for any process of heat supply were equal to dQ/T could not be a function of state, as can be shown by considering the irreversible cycle discussed above. The function would be of very little use, and it would not be entropy!

The essential fact that entropy is a function of state is also often omitted. Sometimes even entropy is defined as 'the amount of heat in a body divided by the absolute temperature'.

Many confused analyses of thermodynamic problems involving entropy are published. Apart from the errors stated above probably the most usual cause of confusion is failure to distinguish between the change of entropy of a particular material and the change of entropy of the whole system.

A common mistake by students is to suppose that changes of entropy are defined for *isothermal* processes only, instead of for *reversible* ones. This may originate from the treatment which is sometimes given in which the idea of a reversible process is very briefly discussed and is confused with that of an isothermal process.

Chapter 7

HEAT TRANSFER

Thermal Conductivity

The rate of conduction of heat (dQ/dt) through a surface of area A in a material is given by the equation

$$\frac{dQ}{dt} = -KA\frac{dT}{dx} \qquad \ldots \ldots (7.1)$$

where dT/dx is the temperature gradient, measured in the direction normal to the surface. The negative sign indicates that heat flows towards parts of lower temperature. This equation is the *definition* of the quantity K which is called the thermal conductivity of the material. It does not by itself express any law of heat flow, since K is otherwise undefined and might *a priori* be expected to be any function of the various parameters.

Both experiments and theories of thermal conduction give the following laws:

(*a*) K is independent of the temperature gradient, hence other things being equal the heat flow is proportional to (dT/dx).

(*b*) Under most conditions K is independent of the area A. At very low temperatures K becomes a function of A, tending to the form 'K is proportional to the square root of A as T tends to absolute zero' for many dielectrics.

(*c*) The value of K is the same for both steady-state and changing conditions. In nearly all elementary courses it is taught that the value of K applies to steady state only, but on going on to problems of non-steady heat flow the same value is used without further discussion.

(*d*) The thermal conductivity of a material is a function of temperature. For metals the value changes very slowly over a wide range of temperature, except at very low temperatures when K tends to zero as T tends to $0°K$ for all materials. Alloys, glasses and gases have thermal conductivities which rise with rising temperature. The conductivities for crystalline dielectrics and some semiconductors rise as the temperature falls, reaching peak values typically at liquid air temperatures. Below this the value of K falls off rapidly.

In elementary courses it is usual to define thermal conductivity as the heat conducted per second between opposite faces of a cube with sides of unit length when the temperature difference is one degree. This is one of many meaningless slogans which students have to learn by heart. It is useless, since for a student to understand and use the idea of thermal conductivity he must learn equation (7.1) and the associated laws, whilst the unit cube 'definition' contributes nothing to this. Such a 'definition' in fact defines nothing whatever, since there is an infinity of different functions which will give the same results for the particular case of the unit cube but different results for all other cases.

Thermal conductivity is sometimes defined to be the 'constant' in equation (7.1), but it should be clear that K is in fact generally not a constant.

In elementary courses it is almost invariably taught that all non-metals are poor conductors of heat compared with metals. In fact many solid non-metals are excellent conductors of heat, especially at low temperatures. At liquid air temperatures some of these materials (*e.g.* synthetic sapphire, Al_2O_3) are better thermal conductors than copper, whilst at room temperatures many non-metals are still far better than many metals. One reason for this very common mistake may be the fact that non-metals which are good conductors of heat are not normally available cheaply as specimens large enough for ordinary school laboratory use. Another cause is probably the widespread misunderstanding of the mechanism of heat conduction. In metals and alloys heat is usually conducted almost entirely by the free electrons. Because of this there is a very close correlation between the electrical and thermal conductivities of these materials which is expressed by the well known approximate law of Wiedemann-Franz-Lorentz. Unfortunately it is often taught that this law applies generally for solids, although actually it does not apply to non-metals. In electrical insulators and semiconductors heat is transferred mainly by extremely high frequency elastic waves quantised as 'phonons'. When heat is transferred by this process there is no relationship between the electrical and thermal conductivities. For example, at a particular temperature the thermal conductivities of specimens of aluminium and sapphire are equal whilst the ratio of their electrical conductivities is of the order of 10^{22}.

The conduction of heat by phonons must occur in all solids, but in metals this effect is greatly reduced as a result of the scattering of phonons by free electrons. Generally the thermal conductivity of a solid is the sum of two parts, one resulting from phonon conduction and the other from electron conduction. Usually one or the other effect is predominant, but in a few cases their contributions are comparable.

At sixth-form level it is almost always asserted that liquids are all poor conductors of heat, although mercury is sometimes named as an exception. In fact liquid metals are all reasonably good conductors of heat. The thermal conductivity of liquid metals is of obvious significance in such processes as casting, welding and soldering. The particularly high values for the molten alkali metals are of importance in nuclear engineering, for example the Dounreay Fast Reactor uses liquid sodium-potassium alloy as its coolant.

It may be remarked that below the λ-transition liquid helium has an incomparably higher thermal conductivity than any other substance, liquid or solid.

Convective Heat Transfer

When heat flows from the surface of a solid to a cooler ambient liquid or gas there is a temperature gradient in the layers of fluid near to the surface. When natural or forced convection currents are free to circulate, the bulk of the fluid at a distance from the hot surface will be at a practically uniform temperature. Thus there will be a temperature difference t between the surface and the bulk fluid. Similar arguments of course apply in the reverse case of a cool surface heated by a warmer fluid.

It is usual to consider q, the heat flow from unit area of the surface in unit time. This is clearly a function of t and of the properties of the fluid. It is also found to be a slowly varying function of the dimensions of the surface, hence the traditional assumption that the total heat flow is proportional to the surface area is incorrect. For forced convection q is naturally a function of the speed of the forced flow.

For strongly forced convection it is found that q is proportional to the excess temperature t over a wide range of values. This is sometimes called Newton's law of cooling. For free convection q is proportional to a higher power of t, usually between 1·25 and 1·5. In this case Newton's law can be used only as an approximation for small values of t.

For many purposes it is very convenient to express results in terms of a quantity denoted by h which is the quotient of q divided by t. This determines the rate of loss of heat from unit area for a given excess temperature. If Newton's law is applicable h will be a constant for given conditions, but otherwise h is a function of t. The quantity h is most suitably called the *heat transfer coefficient,* but in elementary work it is often called *emissivity*. This latter name is used with a very different meaning in radiation theory, and it is most unfortunate that the same term is used for two different quantities in fields which are sufficiently closely connected for considerable confusion to result.

Verification of Newton's Law of Cooling

This is a very common laboratory exercise for students. In one form of the experiment hot water is poured into a can provided with a stirrer and a thermometer, and the temperature is measured as a function of time. From the slope of a cooling curve the rate of fall of temperature is found for various excess temperatures. A graph of rate of cooling against excess temperature is plotted as in *Figure 10*.

The student is told to draw the best straight line through the points. He then concludes that 'the graph is a straight line, showing that the rate of cooling is proportional to the excess temperature, hence Newton's law of cooling has been verified'. This is a very common type of mistake as it is not generally appreciated that for proportionality the graph must be a straight line *through the origin*. In the present case the fact that the best straight line almost always gives a positive intercept on the temperature axis should be a warning.

Clearly the curve must go through the origin since there should be no cooling with zero excess temperature, hence a straight line cannot properly be drawn.

Experiments of this kind can be very instructive, but only if the student is taught to interpret the results intelligently. Newton's law is seldom accurately applicable, partly because the convection is usually either entirely or partially natural, and partly because of the effects of evaporation. In this connection it is important to note that in the teaching of heat

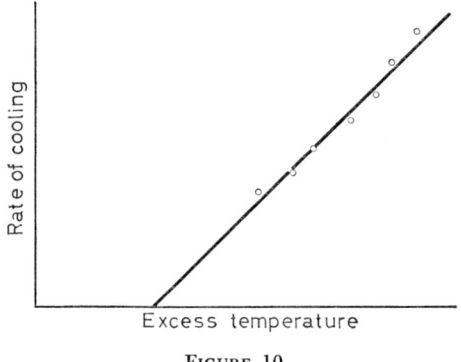

FIGURE 10.

transfer the effects of evaporation and condensation are usually ignored, although in many circumstances they play a dominant part.

Problems in Convective Heat Transfer

As we have shown, there must be a temperature difference t between a solid surface and the bulk of a fluid when heat flows between them. This fact is commonly taught when Newton's law of cooling is discussed, but it is almost invariably totally ignored in all other problems on heat transfer. In a very large number of problems and experiments students are taught to assume that the temperature of a surface is the same as the temperature of the ambient fluid even though there is a large heat flow.

A common type of problem in this field is the calculation of the heat flow through a window. In this case typically a few hundredths of the temperature drop takes place in the glass and the rest takes place in the surface layers of air, hence the traditional method of solving such problems gives a heat transfer typically fifty times too large. Another similar problem is the calculation of the rate of growth of a layer of ice on a pond. Again it is wrongly assumed that the surface of the ice is at the same temperature as the air above it. Many similar problems are set in examinations and given in textbooks. They reach the depths of absurdity when the sheet of material is a metal, for in this case the traditional methods give heat flows thousands of times those obtainable in reality.

With the higher values of the heat transfer coefficient given by liquids the errors are less than in the case of a gas, but even so calculated values of heat flow may be an order of magnitude too large when using the absurd assumptions of school 'physics'.

There is a traditional laboratory experiment to measure the thermal conductivity of a specimen of glass in the form of a tube. The tube is heated on the outside by steam and cooled on the inside by water, the rate of conduction of heat being determined by the rise of temperature of the water. To calculate the conductivity of the glass it is assumed that the outer surface is at the temperature of the steam whilst the inner surface is at the mean temperature of the water. These assumptions are of course invalid even as an approximation, hence the experiment is fundamentally unsound. The author has been told by one G.C.E. examination board that *the experiment has been used in teaching laboratories for at least fifty years, so there must be good educational reasons for doing it.* It is hoped that comment is unnecessary.

CHAPTER 8

ELECTROMAGNETIC RADIATION

Heat and Light

The radiation from a hot body can be dispersed by a spectrometer to form a spectrum. If a very sensitive thermometer is moved through the spectrum the intensity of radiation can be found as a function of wavelength by the heating effect. Obviously all parts of the spectrum detectable by such means equally merit the name 'heat'. For solar radiation the maximum heating effect in a given range of wavelengths is found at a wavelength roughly 0·5 microns. Between wavelengths of about 0·4 microns and 0·7 microns the radiation excites vision in typical human subjects and is thus called light. The limits of the visible spectrum are not well defined. For a given observer the limits change according to the level of illumination, moving to shorter wavelengths at lower intensities (the Purkinje effect). The limits differ for different individuals, and many unfortunate persons lack any effective selective detectors for these wavelengths and are said to be *blind*. Such individuals can, just as can persons with sight, detect these and similar radiations by the heating effect on their skins. Contrary to what is often taught light can be *felt*.

The thermal radiations of shorter wavelengths than the visible are called ultra-violet and the longer ones infra-red. As the limits of the spectrum are so ill-defined it is not possible to state exactly the proportions of energy in each division, but for solar radiation roughly forty per cent of the heat is in the visible region, about an eighth is ultra-violet and the rest is infra-red. For lower temperature sources a much larger proportion of the heat is radiated in the infra-red. It is probably for this reason that in elementary teaching radiated heat is usually identified with infra-red radiation. This error leads to the absurd idea that light has no heating effect. The part of the spectrum which is visible depends on the eyes of the observer, hence it appears that many physics teachers believe that the existence of eyes with sensitivity over a given range of wavelengths magically destroys the heating effect of this radiation.

The confused teaching on the relationship between radiant heat and light is exemplified by such statements as 'radiant heat, that is infra-red, can be detected with a thermopile, and light can be detected with a photoelectric cell'. It should instead be taught that all parts of the spectrum can be detected by thermometers such as the thermopile, whilst there are photoelectric cells for each part of the spectrum. Light is just a part of the spectrum of radiation selected by a particular detector, and what is visible to one animal may be invisible to another.

Short-wavelength Radiations

There is much confusion in the terminology of the radiations of much shorter wavelength, and this confusion is not confined to elementary teaching.

The energy of a quantum of radiation of wavelength λ is equal to hc/λ where h is Planck's constant and c is the speed of electromagnetic radiation in a vacuum. Approximately

$$h = 4 \cdot 1 \times 10^{-15} \text{ electron-volt secs. (eV sec.)}$$

and $c = 3 \cdot 0 \times 10^8$ m.sec.$^{-1}$

Hence for radiation of wavelength one Angstrom unit (10^{-8} cm) the quantum energy is $12 \cdot 3$ keV.

Between the near ultra-violet and quantum energies of a few keV electromagnetic radiations are very strongly absorbed in matter by the photoelectric effect and are usually studied inside vacuum apparatus. In this section we are concerned with those radiations with wavelengths less than a few Angstroms, that is with quantum energies above a few keV. These radiations are penetrating and can be studied without difficulty.

In elementary teaching it is usual to divide the short-wavelength end of the spectrum into X-rays, γ-rays, and cosmic rays. The divisions between these supposedly separate parts of the spectrum are given various fanciful values. Thus according to some authors X-rays have quantum energies up to 10^4 eV, and according to others up to 10^6 eV. Gamma rays are supposed to have energies greater than X-rays, and radiations of still shorter wavelength are called cosmic rays. Typically γ-rays are said to have wavelengths of 10^{-10} to 10^{-11} cm, that is quantum energies 1·2 to 12 MeV.

It is necessary to clarify this subject. There are two main classes of radiations known as X-rays. There is a line spectrum of X-rays characteristic of a given element emitted by atoms which have lost electrons from inner shells. Atoms are usually so ionized by bombardment with high energy electrons in X-ray tubes, but there are other processes by which inner electrons may be removed, for example by nuclear capture. In any case the wavelengths are determined by the atomic number. The highest quantum energies of characteristic X-rays are of roughly 100,000 eV and are emitted by elements with high atomic number such as uranium. X-rays are also emitted when unbound electrons are accelerated on passing near to atoms. This gives a continuous spectrum with quantum energies up to a maximum which is very nearly equal to the kinetic energy of the incident electron. Thus the only lower limit which can be set to the wavelength of this radiation is determined by the machines available to accelerate electrons. At present accelerators exist which can give electron energies up to 10^9 eV, hence electromagnetic radiation can be generated with quantum energies up to this value. Such energies are so much greater than those obtained with ordinary medical and industrial X-ray equipment that many physicists use the term *bremsstrahlung* in preference to X-radiation. This is not very helpful since this name refers to the mechanism of production and is equally applicable to the continuous X-rays from low energy machines. There is no natural division since machines work at all intermediate energies, hence it is quite reasonable to say that X-rays have energies up to 10^9 eV.

The name gamma radiation was originally given to electromagnetic emission from radioactive materials to distinguish it from the electrically charged α- and β-rays. If we retain this meaning the range of energies is far less than that for the continuous X-ray spectrum. There is probably no real definable long wavelength limit, but the rays usually studied have energies above about 10 keV. The discovery of the Mössbauer effect has given great importance to certain low energy γ-rays, particularly the 14·4 keV radiation from an excited state of Fe^{57}, and there is no excuse for teaching that γ-rays must have energies of millions of electron volts. Actually there are very few nuclides emitting radiations with quantum energy much above 2 MeV and hardly any above 6 MeV.

In the early days of atomic physics few X-ray tubes operated above a few tens of kilovolts, whilst the γ-rays which were most studied were generally those of high energy, hence the idea arose that γ-rays have wavelengths less than those of X-rays. This has led to an unfortunate confusion in terminology, since some writers use the terms X- and γ-rays to distinguish energy ranges, rather than radiations from different types of source.

There are sources of high energy quanta other than electron accelerating machines and radioactive nuclei. When electrons combine with positrons, quanta are produced most of which have energies of 0·51 MeV. This is best called *annihilation radiation*. Much higher energy quanta are produced by the decay of heavy unstable particles such as neutral pions and neutral sigma hyperons. These various radiations are sometimes called γ-rays although they do not originate in excited states of nuclei.

It may be that the detection of these latter types of radiation, and also bremsstrahlung, in secondary cosmic radiation has helped to perpetuate the common mistake of supposing that cosmic rays are extremely short wavelength electromagnetic radiations. In fact it has been known for many years that the primary cosmic rays are nuclei, mostly protons, and that the penetrating secondary radiations are mostly muons. The properties of both the primary and secondary particles are so completely different from those of electromagnetic waves that there can be no justification for teaching that cosmic rays are electromagnetic.

Optical Spectra

According to some traditional sixth-form courses there are

four types of optical spectra, namely line, band, continuous and absorption. This extraordinary classification has been explicitly demanded by G.C.E. examiners.

The classification of spectra into four types is misleading. It implies that an absorption spectrum is neither line, band nor continuous, whereas it will of course be one or more of these. Further it implies wrongly that line, band and continuous spectra cannot be obtained by absorption. These absurdities can be contrasted with the simple facts of spectroscopy. A spectrum is normally obtained either by emission or by absorption. Both classes may be either line, band or continuous, hence there are *six* types of spectra.

It is traditionally taught that continuous emission spectra are obtained with light from incandescent solids whilst line and band spectra are given by gaseous sources. Sources of line spectra are sometimes correctly stated to be atoms, and sometimes incorrectly said to be elements. Similarly band spectra are attributed correctly to molecules or incorrectly to compounds. Such accounts of the origin of spectra leave much to be desired even at their best. They raise the obvious question, what type of spectrum is given by the light from a hot liquid? Further, anyone who has seen the spectrum of sunlight knows that it is primarily continuous. Since it is taught that continuous spectra originate from solid sources this suggests that the sun is solid, although as is well known it is gaseous.

The essential facts about the origin of spectra are actually quite simple. In gases at low pressures the atoms or molecules are widely separated and have well defined energy levels. Transitions between various bound states of atoms give rise to emission and absorption lines. Molecules give rise to band spectra as a result of changes of rotational or vibrational energy levels occurring simultaneously with the changes of electronic states. There is a continuum of energies of unbound states of electrons interacting with ions which gives weak continuous spectra. Generally in an electric discharge or flame there will be molecules, separated atoms and ions present, giving lines and bands with a faint continuous background in the emission spectrum. At high pressures the atoms and molecules cannot be regarded as isolated free particles and their energy levels are perturbed by interactions with their neighbours, giving pressure broadening of the spectral lines. In solids and liquids, and also gases at very high pressures, the molecules are so close as to interact continually. Hence the optical emission spectra of these dense materials are continuous.

CHAPTER 9

PRINCIPLES OF CURRENT ELECTRICITY

Potential Difference and Electromotive Force

These quantities are of great importance in the study of electric circuits. The author has found that only very few university freshmen have any clear ideas of their meaning. Generally speaking P.D. is known purely as a synonym for 'voltage', and most students consider that E.M.F. is just another name for the same thing. Usually students know that the E.M.F. of a cell is equal to the P.D. between the terminals when an open circuit, but as they rarely know what either P.D. or E.M.F. means this is not very helpful.

There is certainly great confusion concerning the meanings of these terms in the traditional teaching of electricity, and some authors and many teachers have obviously cut the Gordian knot by avoiding any explicit discussion of the subject. It is undoubtedly very difficult, and it is with considerable diffidence that the author presents this exposition in the hope that it may clarify some points.

In electrostatics the idea of potential difference is fairly simple. If a very small charge q is moved between two points the work done divided by q is called the P.D. It is necessary to specify a *very small* charge, as otherwise its presence and motion will alter the quantity which is to be determined. The common reference to 'unit charge' in this context is unhelpful.

In a purely electrostatic problem there is normally no problem of the dissipation of energy. Generally if a charge is moved mechanically against the field the work done is stored as energy in the field and can in principle be recovered on allowing the charge to retrace its path. In current electricity the problem is different. Consider a part of a wire carrying a current. For simplicity let the wire be of uniform composition and temperature, and be placed in a vacuum. A field of force must be acting along the wire to drive the current. This field will exist not only inside the material but also in the surrounding space, hence there is an electrostatic difference of potential between points just outside the wire at either end of the part considered. The work done by the field in moving a very small charge from one end to the other inside the wire is the same as that done in moving an equal charge through the surrounding space between points just outside the ends of the wire. Hence the current electrical P.D. within the wire is in this case equal in magnitude to the electrostatic P.D. outside it. Despite the equality of magnitude the two quantities are not identical in character. The energy used in moving a charge in a vacuum is in principle recoverable unless it hits a body, but the energy used in moving a charge along a wire is dissipated in the form of heat. It is therefore desirable to define P.D. in current electricity in terms of the dissipation of energy, that is, the P.D. across a conductor carrying a steady current is the rate of heat production divided by the current. The M.K.S. unit of heat is the Joule and that of rate of heating is the watt. The coulomb is the unit of charge and the ampere of current, hence the M.K.S. unit of P.D. is the volt, where

$$\text{volts} = \frac{\text{joules}}{\text{coulombs}} = \frac{\text{watts}}{\text{amps}}$$

The volt is equally suitable as a unit of electrostatic P.D. and current electrical P.D. The fact that the same unit is used for two different types of quantity should cause no difficulty. We are familiar with the joule as a unit of both kinetic and potential energy but we do not suppose that the two kinds of energy are identical.

If heat is being generated there must be some source of energy. Anything which can supply energy to an electric current is said to have an E.M.F. The E.M.F. is defined as the power supplied divided by the current driven, and the M.K.S. unit is the volt.

The essential differences between electrostatic P.D., current P.D. and E.M.F. can be shown by considering the reversal of the direction of motion of charge. In the electrostatic case the charge moved must be too small to affect the P.D. significantly. If a positive charge is allowed to move down the potential gradient work will be done on it by the field. If the charge is forced mechanically in the reverse direction energy is stored in the field. In current electricity if the direction of current flow is reversed the P.D. is reversed, but the rate of energy dissipation is unchanged. It is impossible to convert heat back into other forms of energy by sending the current backwards through the resistance. An E.M.F. does not involve dissipative forces. If an E.M.F. drives a current then it gives energy to the current. If the current is reversed then the current must give energy to the source of E.M.F. This case is very similar to the electrostatic one, and in fact in any closed loop the E.M.F. is the line integral of the electric field around the loop.

An electromotive force can arise in several ways, in each of which electrical energy can be put into or taken from a 'store'. For example, in an accumulator the cell gives energy to the current when this flows in the direction of the E.M.F. and if the current is forced in the opposite direction the cell is charged, the current then giving energy to the cell. In practice there are always some dissipative effects and no cell is perfectly reversible. Similar complications arise in all circuit elements, for example an inductor normally has resistance as well as inductance, but these will be ignored in our discussion.

Energy can be stored in a magnetic field, the energy per unit volume being $\frac{1}{2}$ BH (in rationalized units). The electromotive force produced by changing the contents of a 'magnetic store' is called an induced E.M.F. When a current in a circuit decreases energy is extracted from the magnetic field and the E.M.F. acts in the direction of the current. When a current increases the E.M.F. is in the opposite direction to the current so work is done by the current to store energy in the field.

Energy can similarly be stored in an electric field, particularly in a capacitor. When a capacitor is being charged the E.M.F. opposes the current thereby transferring energy to the electric field in the dielectric. On discharge the E.M.F. of the capacitor acts in the direction of the current giving it energy at the expense of the field.

Instruments for measuring 'voltages' in effect measure the electrostatic P.D. between the terminals to which they are connected. Consider a voltmeter connected across a resistor and a cell in series forming part of a circuit in which a current i flows. For a resistance R and electromotive force E the reading of the voltmeter may be $(Ri+E)$ or $(Ri-E)$ according to which way round the cell is connected. Thus generally the meter does not read either the current electrical P.D., which is Ri, or the E.M.F., although in certain limiting cases it may measure either. If the cell and resistor are placed in a calorimeter the current electrical P.D. can be measured by the heating effect. The results obtained by this method will clearly not agree in general with those given by a voltmeter.

It is important to note that since an E.M.F. is the line integral of an electric field a current does not have to flow for an E.M.F. to be generated. We have referred to a current in the definition, but in principle this current can be allowed to tend to zero, as in the case of a cell on open circuit. This is particularly important in the case of a magnetically induced E.M.F. which can arise in any medium, including a vacuum. The latter case is one of the foundations of the electromagnetic theory of radiation, and is essential for the operation of devices such as betatrons. Unfortunately it is customary to teach that an induced E.M.F. can only arise in a conductor. For example, examination questions are set in which the candidate is asked to calculate the E.M.F. generated between the wing-tips of an aircraft flying in the earth's magnetic field *assuming the wings to be conductors of electricity.*

Resistance and Ohm's Law

Let the current-electrical P.D. across a body be V and the current i. A very important property of the body is then its resistance R *defined* by

$$R = \frac{V}{i} \qquad \ldots \ldots (9.1)$$

Since the quantity R has no other definition this equation does not express any law of nature, but just gives a name to the quotient of V divided by i.

For certain devices we are mostly concerned with changes of current resulting from the potential difference altering from a normal reference value. In this case the name resistance will be given to $\delta V/\delta i$. We shall, however, limit our discussion to the resistance of a simple conductor, defined by equation (9.1).

For certain conductors, providing the temperature is constant, the resistance is independent of the current. This result is Ohm's law, which can be expressed as

$$R = \text{Constant, if } T \text{ is constant.}$$

This law is obeyed very exactly indeed by metals and alloys and approximately by several non-metallic materials. Some 'non-ohmic' materials deviate very markedly from Ohm's law, such deviations being of great practical importance in some cases. If a metal wire carries a relatively heavy current the temperature will be raised and hence the resistance is a function of the current. In these circumstances although the material is ohmic in character Ohm's law is not applicable.

In the teaching of electricity these simple and vital ideas are almost invariably confused. It is usual to state that at constant temperature for any conductor Ohm's law is obeyed, that is that V/i is a constant, and that this constant is called the resistance R. Thus equation (9.1) is said to express Ohm's law. The logical weakness of this approach should be apparent, since it leaves resistance undefinable in all circumstances in which Ohm's law is not applicable. Consider a simple experiment in which a metal-filament lamp is connected in series with a battery, a variable resistance and an ammeter, whilst a voltmeter is connected across the lamp. The current through the lamp is varied over a wide range and corresponding readings of i and V are recorded. From the results it is apparent that V/i is not constant. If the subject were taught logically there would be no difficulty in interpretation. The resistance could be evaluated for each current, and it would be observed that R increased with increasing i. As the temperature of the filament is obviously higher for higher currents it would be concluded that the resistance of a metal wire increases with increasing temperature. According to the traditional way of teaching this subject the experiment cannot be interpreted. The definition of resistance is made to depend upon Ohm's law, which is inapplicable when the temperature varies. Hence the ideas of resistance and of temperature coefficient of resistance cannot be used in this simple case, which is preposterous. Elementary textbooks do sometimes describe this experiment, and in this context it is said the V/i is the resistance of the filament. However, such books explicitly define resistance as the constant in Ohm's law. The contradiction should be obvious.

Since a direct potential difference is a simple scalar quantity the potential difference across a number of resistances in series must be the sum of the individual values. Hence the resultant resistance is just the sum of the separate resistances. This follows directly from the *definition* of resistance and contrary to the usual teaching does not depend in any way upon Ohm's law. Resistances in series give a resultant equal to the sum of the individual values even if they are functions of the current, although of course one must consider the values for the particular current. One important corollary is that the principle of the potentiometer does not depend upon Ohm's law.

By contrast the reciprocal rule for resistances in parallel does depend upon Ohm's law since the current generally divides unequally between the branches.

It seems almost incredible that such distinct ideas as the definition of a quantity and the statement of a natural law could be confused, but the confusion is in this case almost universal. Intelligent students are bewildered when they learn of 'non-ohmic' resistances, since according to the way they have been taught this is a contradiction in terms.

One consequence of the traditional muddled approach is that students rarely understand any of the implications of Ohm's law. In many cases it is firmly believed that V/i must be constant in all circumstances, hence the students' minds are closed against the recognition of any other result. Another common misunderstanding is just the opposite of this, namely the supposition that Ohm's law does not apply when in fact it does. This is illustrated by students' reactions to an experiment in which the resistances of sheets of uniform gauge metal are measured by the ammeter-voltmeter method. If pairs of terminals are fitted at equal separations on various sheets of similar material of different width it is found that the resistances are not inversely proportional to the widths of the specimens, in consequence of the lines of current flow not being parallel near the terminals. This result has of course no relation to the validity of Ohm's law, but many students assume that it proves that the law does not hold.

Verification of Ohm's Law

The methods for verifying Ohm's law which are usually taught are unsound, and the arguments employed to justify them are unconvincing. There is firstly considerable misunderstanding as to the nature of the law. It is not a universal fundamental law of nature such as, for example, the law of conservation of momentum, but is a law describing the behaviour of a certain property of materials under particular conditions. Thus we cannot examine Ohm's law in general, but can only examine particular materials to determine whether or not they conform to it. If it is shown that copper, for example, obeys the law very exactly, this does not prove that silicon carbide will do so. Similarly the failure of one material to obey the law does not invalidate its application in another case.

The most obvious approximate way of testing Ohm's law for a particular conductor is to use the ordinary ammeter-voltmeter method to measure the resistance for various values of the current. This approach is almost universally condemned in elementary teaching on the grounds that the calibration of the moving-coil meters depends upon the law one is testing. The question of calibration must be considered further, but the immediate point to note is that this objection is generally irrelevant, since the materials used in the meters are normally different from the material under test.

To avoid the use of moving-coil meters it is usual to recommend the employment of certain supposedly absolute meters such as electrostatic voltmeters or quadrant electrometers to measure the potential difference. Now such instruments are not in fact absolute but require calibration against a potentiometer just like moving-coil voltmeters, hence they have no advantage for this purpose. As the current through the coils of a potentiometer is maintained constant in use Ohm's law is irrelevant, and this instrument or a voltmeter calibrated against it can in principle be used to verify the law.

An ammeter is normally calibrated by passing a current through it and a standard resistance in series, and measuring the P.D. across the resistance with a potentiometer. The resistance must remain constant, hence it must obey Ohm's law and be maintained at constant temperature. Thus it is only if Ohm's law is known to be true for the standard resistance that the ammeter can be used to test its validity in other cases. It is interesting to note that it is usually the calibration of the voltmeter rather than that of the ammeter which is assumed to depend upon Ohm's law.

In elementary courses various crude methods of testing Ohm's law are described, mostly requiring the use of electrostatic instruments. These methods are often unsound in principle, but even when they are otherwise suitable they are imprecise and can only be applied over a very limited range of currents. For null experiments it is necessary to establish the law for current over a range of several orders of magnitude, and to a precision at least a hundredfold better than for work using pointer instruments. Obviously only null methods are suitable to test the law adequately. As such methods appear to be little known one is described briefly below.

A Wheatstone bridge circuit is used in which two adjacent arms consist of identical resistance coils. These coils are constructed of equal lengths of the same wire, such that even if the material used does not obey Ohm's law their resistances will remain equal provided they carry equal currents. The other two arms of the bridge consist of wires of the material under test. One wire is long and thick, whilst the other is short and thin. With a small current flowing the bridge is balanced by adjusting the lengths of these wires, hence their resistances are equal for small currents. The current in the circuit is now increased by several orders of magnitude by shorting a resistor in series with the battery. If Ohm's law is obeyed by the material under test the bridge will remain balanced initially, although heating effects may quickly disturb the balance. If the material is non-ohmic the thin wire will be affected by the current more than the thick one, since the current per unit area in the material is greater. Hence balance will be destroyed. It can be seen that if resistance were directly proportional to the current the bridge would still remain balanced, but such exceptional possibilities are easily excluded by simple tests.

Effective and R.M.S. Values

It is generally taught that the effective value of an alternating current or potential difference is the root mean square value (R.M.S.). This is only true if the resistance is constant, and hence does not apply to a non-ohmic conductor, or to a filament the temperature of which is a function of the instantaneous value of the current.

If the instantaneous value of the current is i and that of the P.D. is V, then the power W is given by

$$W = \overline{(iV)}$$

where the product of the instantaneous values is averaged over a cycle. Since $V = iR$ by definition we have

$$W = \overline{(i^2 R)} = \left(\overline{\frac{V^2}{R}}\right)$$

Now if R is a constant clearly $W = Ri^2$, but if R varies throughout the cycle

$$W \neq \bar{R} \times \overline{i^2}$$

since the average of a product is not generally equal to the product of the averages of the factors.

Inductance

When an electric current changes there is a resultant change of the energy of the associated magnetic field causing an induced E.M.F. For any component in a circuit the magnetically induced E.M.F. divided by the rate of change of

current is called the self inductance, usually denoted by L.
Thus the induced electromotive force E is given by

$$E = L\frac{di}{dt}$$

Very commonly there is ferromagnetic material present, in which case the value of L is a complicated function of the current. Hence the E.M.F. is *not* generally proportional to the rate of change of current as is usually taught. It is thus incorrect to follow the common practices of defining L as a constant of proportionality, or stating that it is the induced E.M.F. for unit rate of change of current.

Similar remarks apply to mutual inductance.

The Magnetic Circuit

There is a well-known analogy between electric current and magnetic flux ϕ, that is the product of area A and the normal component of induction B. On moving a distance l along the lines of force we move through a magnetic potential difference U equal to field multiplied by distance, i.e.

$$U = Hl = \frac{Bl}{\mu} = \phi \times \frac{l}{\mu A} \qquad \ldots \ldots (9.2)$$

This equation is analogous to the electrical one

$$V = iR = i \times \frac{l}{\sigma A}$$

where σ is the conductivity.

The quantity $(l/\mu A)$ is the analogue of resistance and is called reluctance. The magnetic flux is maintained by a magnetomotive force provided by coils or by magnetic poles.

Since the important applications of this concept are to magnets with ferromagnetic cores the permeability μ is a variable, being a function of ϕ. This is similar to the case of a non-ohmic resistance, for which the conductivity σ is a function of the current.

It is usual to describe equation (9.2) as the analogue of Ohm's law, which is obviously incorrect. If a part of a magnetic circuit does not contain ferromagnetic material then the reluctance of this part is a constant, being independent of the flux. It is this last statement which is analogous to Ohm's law.

Chapter 10

C.G.S. UNITS OF ELECTRICITY AND MAGNETISM

Introduction

There are two commonly used systems of electrical units based on C.G.S. mechanical units. These two systems are used haphazardly, often without the user being aware which one he is using, or even of the fact that there are two systems. They differ in the definitions and in the dimensions of certain quantities.

System I

This is based upon the inverse square laws of force between point charges and point poles written in the following forms

For charges q_1 and q_2, force $= \dfrac{q_1 q_2}{\kappa r^2}$ . . . . (10.1)

For poles ϕ_1 and ϕ_2, force $= \dfrac{\phi_1 \phi_2}{\mu r^2}$ . . . . (10.2)

In these equations r is the distance between the pairs of point charges or poles and μ and κ are magnetic and electric properties of the media called permeability and permittivity respectively. For empty space μ is given the value one electromagnetic unit of permeability, and κ is given the value one electrostatic unit of permittivity. The dimensions of q and ϕ are then expressible in terms of the dimensions of κ and μ as well as mass (M), length (L) and time (T).

Thus the dimensions of electric charge are

$$[q] = [M^{1/2} L^{3/2} T^{-1} \kappa^{1/2}]$$

And the dimensions of pole are

$$[\phi] = [M^{1/2} L^{3/2} T^{-1} \mu^{1/2}]$$

Those units which explicitly involve the dimensions of κ are called electrostatic units (e.s.u.) whilst those which explicitly involve the dimensions of μ are called electromagnetic units (e.m.u.).

If a very small pole is placed in a magnetic field of force it will be subjected to a force. This force divided by the magnitude of the pole is called the magnetic field intensity and is denoted by H. Since isolated poles do not exist this definition has been criticized. To be precise we could instead consider the couple on a dipole, since dipoles certainly do exist, but it is easier to think in terms of the separated pole and we shall lose nothing by doing so. The dimensions of the magnetic field intensity are given by

$$[H] = [\text{force} \div \text{pole}]$$
$$\therefore \quad [H] = [M^{\frac{1}{2}} L^{-\frac{1}{2}} T^{-1} \mu^{-\frac{1}{2}}]$$

In this system the unit of field intensity is sometimes called the *oersted* and sometimes the *gauss*.

An electric current produces a magnetic field. If a small element of a circuit carrying a current i subtends an element of angle $d\alpha$ at a point P at a distance r from the current element, the field intensity at P caused by this element is dH_p, where

$$dH_p = \frac{id\alpha}{r}$$

This equation defines the unit of current and determines the dimensions of i as those of field times length.

Hence $\quad\quad\quad [i] = [M^{\frac{1}{2}} L^{\frac{1}{2}} T^{-1} \mu^{-\frac{1}{2}}] \quad\quad$ .... (10.3)

A pole of strength ϕ placed at P will experience a force $\phi i d\alpha / r$ because of the current element. By Newton's third law the element of the circuit will be subjected to an equal and opposite force. The mechanism giving rise to this force is clearly the interaction of the current element with the magnetic field of the pole ϕ. This field will be $\phi/\mu r^2$ at the position of the element, whence it is easily deduced that the force on a conductor carrying a current i in a magnetic field of intensity H is given by

$$\text{Force} = \mu H i l = B i l$$

where l is the component of the length of the conductor perpendicular to the direction of the field.

The quantity B is called the magnetic induction, or magnetic flux density. The dimensions are given by

$$[B] = [\mu H] = [M^{\frac{1}{2}} L^{-\frac{1}{2}} T^{-1} \mu^{\frac{1}{2}}]$$

In this system the unit of induction is usually called the *gauss*.

For generations students have struggled with a quite unnecessary confusion between the distinct functions H and B. It should be clear from this analysis that the field intensity H determines the force of the field upon magnetized materials, whilst the induction B determines the force of the fields upon circuits carrying currents and upon moving charges. In a vacuum the numerical value of H in electromagnetic units of field intensity is equal to the numerical value of B in e.m.u. of induction. This does not mean that H is equal to B, which is a dimensionally distinct quantity, any more than ten yards are equal to ten minutes.

In ferromagnetic materials, and in certain circumstances in other media, the permeability is a function of the field intensity, hence it is not then correct to teach that B is proportional to H.

Electric current has been defined above in terms of its magnetic effect. It can also be considered as a rate of flow of charge, hence we can define an electrostatic unit of current as a rate of flow of one e.s.u. of charge per second. The dimensions of current are thus seen to be given by

$$[i] = [qT^{-1}] = [M^{1/2} L^{3/2} T^{-2} \kappa^{1/2}] \quad \ldots \quad (10.4)$$

It is an essential feature of this system of units that a physical quantity always has the same dimensions in whatever units it is measured, just as a distance has dimension length whether it is measured in feet or in metres. Hence the dimensions of current given in equations (10.3) and (10.4) must be identical. This requires that

$$[\mu \kappa] = [L^{-2} T^{2}] = \left[\frac{1}{\text{velocity}^2}\right]$$

As μ and κ are related in this way it is apparent that only one new dimension is required for the description of electrical and magnetic quantities. This could most conveniently be charge or current, as will be familiar to those who know the M.K.S. system of units, which permits a much clearer dimensional analysis of these quantities.

Precise methods exist for the comparison of various electrostatic and electromagnetic units. It is found, for example, that one e.m.u. of current is very nearly 3×10^{10} e.s.u. of current. It is important to note that in this system the conversion factor is a pure number, and is not the speed of light as is often stated.

System II

This also is based upon equations (10.1) and (10.2), but the permeability and permittivity are now defined to be dimensionless quantities. This is in some ways a simplification, but it does lead to certain difficulties.

Electric and magnetic phenomena are interrelated, hence in this system the dimensions of a physical quantity depend upon the units in terms of which it is measured. If a current is measured by its magnetic effects its dimensions are $[M^{\frac{1}{2}}L^{\frac{1}{2}}T^{-1}]$ as is seen on eliminating the dimensions of permeability from equation (10.3). If it is considered as a flow of charge current has the dimensions $[M^{1/2}L^{3/2}T^{-2}]$ given by equation (10.4). Hence the quotient of the current values in e.m.u. and e.s.u. has the dimensions of speed, and is in fact the speed of light in a vacuum. This is analogous to the case where a distance is expressed either in hundreds of yards or in minutes walking, the ratio of the units being a speed.

In System II the dimensions of magnetic field intensity are $[M^{\frac{1}{2}}L^{-\frac{1}{2}}T^{-1}]$ and the unit is sometimes called the gauss and sometimes the oersted. Thus there are three dimensionally distinct units for which there are only two names, neither of which is used consistently.

In this system of units the status of induction is rather obscure. From the simplest point of view it now has the same dimensions as H and the e.m.u. could reasonably have the same name. H and B are still different quantities which in general have different values and even for a vacuum have different significance. The case is analogous to that of the length of a ship. Here there can be defined two quantities, the length overall and the length at the waterline, which have distinct meanings although they are both measured in the same units and may in certain cases be equal in magnitude. This simple approach is not sufficient if the phenomena are not purely magnetostatic. In e.s.u. the permeability is no longer dimensionless and its value for a vacuum is no longer unity. Hence considerable care is needed on considering such questions as the force on a unit charge or the electric displacement caused by an induced E.M.F.

Comments on the Two Systems

Each of these systems has its advocates and some teachers and authors use one and some use the other. Unfortunately many others use a confused mixture of the two. Very commonly κ and μ are given dimensions as in system I whilst the quotient of the units is given dimensions as in system II. In the use of both systems there is great confusion over the meanings of H and B. These difficulties are becoming very serious indeed now that physicists often find it necessary to work both in C.G.S. and in M.K.S. units and to be able to interconvert between them. To avoid difficulty it is essential that the C.G.S. system chosen should be clearly defined and used consistently. In particular it is important to be clear that it is the quantity B and not H which should appear in equations giving an induced E.M.F. or the force on a conductor or a moving charge.

Chapter 11

ATOMIC PHYSICS

Radioactivity

Figure 11 is one variation of one of the most universally taught diagrams in elementary physics. It purports to represent the paths of the radiations from a radioactive substance when a magnetic field is directed downwards through the plane of the figure. No such result could possibly be obtained. For a material which emits both α- and β-rays the radius of curvature of the path of the most energetic β-particles is

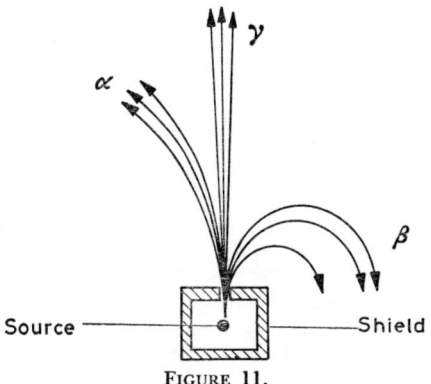

FIGURE 11.

typically a hundredth of that of the α-particles, whilst nearly all the β-particles will follow even more sharply curved paths. If the shield is thick enough and the hole through it is narrow enough to collimate the γ-rays then the β-rays will not pass through it, as a result of their deflection in the fringing magnetic fields. The diagram is very schematic indeed, and in practice is very misleading.

This diagram does appear to have a very strong appeal,

81

and it certainly succeeds in impressing many students with the wrong idea that all radioactive materials emit these three kinds of rays. This wrong idea is in fact sometimes explicitly taught.

Most sixth-form teaching on this subject is many years out of date and is quite unsystematic. For this reason a brief account of radioactivity is given below.

An α-ray is a helium nucleus consisting of two protons and two neutrons, ejected with an energy of a few MeV from the parent nucleus. This form of activity is confined almost entirely to elements of atomic number greater than 82, but there are a few very long-lived α-emitters with atomic numbers about 60. Most α-radioactive materials which are used in significant quantities occur in nature, although one such material of great importance, plutonium 239, is manufactured on a large scale in nuclear reactors.

Negative beta rays are electrons which are emitted with a continuous spectrum of energies up to a maximum value characteristic of the nuclide. This maximum varies from about 10 keV to several MeV. Few beta-active materials are found in significant quantities in nature, but very many can be prepared in the laboratory and most elements have at least one isotope which emits β-rays. This type of activity is characteristic of nuclei which have an excessive proportion of neutrons, and the effect of the process upon the nucleus is the conversion of one of the neutrons into a proton.

A few long-lived β-emitters such as potassium 40 are found in nature, having been produced like the long-lived α-emitters by the cosmic processes which created our earth. Other β-emitters of various lifetimes are produced by the successive emission of α-rays in the natural decay series. Very heavy nuclei contain more neutrons than protons, hence the emission of alpha particles increases the ratio of the number of neutrons to the number of protons, thus a succession of α-decays will produce a nucleus with an excessive proportion of neutrons.

When very heavy nuclei are fissioned the proportion of neutrons in the fragments is greater than the normal value for nuclei of medium size, hence the fission products are nearly always negative beta emitters, and in most cases give several successive decays before becoming stable. This is the cause of most of the dangerous 'fall-out' from nuclear bombs.

Negative beta emitters are most commonly produced by irradiating ordinary matter with neutrons. This is a routine operation in many nuclear reactors.

Positive beta-rays are *positrons,* that is particles with the same mass as electrons but with a positive charge. They are emitted by many nuclides, particularly those of low atomic number, when there is an excessive proportion of protons. This decay process results in the conversion of a proton into a neutron.

Another mode of decay of nuclei with excess of protons is electron capture. A proton is converted into a neutron by the capture of an electron from an inner shell of the atom. One way in which this process may be observed is by the detection of the characteristic X-rays emitted by the resultant excited atoms.

Nuclides with an excessive proportion of protons are commonly produced by bombarding ordinary materials with light nuclei such as alpha particles and deuterons in accelerating machines.

Nuclei with even numbers of protons and neutrons are more stable than those with odd numbers. In consequence there are some nuclides with odd numbers of both types of particle which exhibit both negative and positive beta decay processes.

When a nucleus is excited it may return to the ground state either by emitting a γ-ray or by ejecting an orbital electron from the atom. These electrons can be distinguished from negative beta rays by their line spectrum of energies. Nuclei are often produced in excited states as a result of the various types of radioactive decay discussed above. Thus it is common to find that a substance which emits α- or β-rays also emits γ-rays. There are however many radioactive substances which do not emit γ-rays, as the resultant nuclei in these cases are normally produced in the ground state.

The elementary systematic rules of nuclear stability and types of decay which are outlined above were clearly established *thirty years ago.* In a recent study of a number of sixth-form textbooks the author found no discussion of these ideas. Sometimes it is taught that all radioactive materials emit α-, β- and γ-rays, sometimes that γ-rays are always emitted by α- and β-emitters. Not one of the books examined gives any idea of the types of nuclei which undergo the various modes of decay. Not one describes electron capture or (with one possible exception) positive beta decay. The possible exception is a book in which the positron is confused with the proton so that the nature of the process described is obscure.

Absorption of Radiation in Matter

In elementary work prominence is rightly given to the subject of the absorption of the various types of rays in matter. Unfortunately the treatment given to this important topic is generally very unsatisfactory. There is partial truth in the common generalization that penetration is least for α-rays and highest for γ-rays, but beyond this the information given is usually unreliable.

Alpha-rays typically can penetrate matter of areal density of a few milligrams per square centimetre, and hence will have a range of a few centimetres in air at S.T.P., or about a hundredth of a millimetre of aluminium.

It is usual to teach that β-rays will penetrate one centimetre of aluminium. Whilst it might be true that the fastest electrons from some extremely short-lived nuclides will have such a range, this figure gives a very false picture of the normal properties of such rays. The most penetrating particles from the very energetic daughter of the notorious isotope strontium 90 will penetrate about 3 mm. of aluminium, but by far the greater part of the radiation is absorbed in a much thinner layer. The rays from a medium-energy β-emitter such as carbon 14 are fully absorbed by a thick sheet of paper or a few inches of air. Very low energy β-rays, such as those from tritium, are less penetrating than α-rays. To study most beta radiations, detectors such as Geiger counters must have a thin 'window', usually of mica, to admit the rays.

A very wide range of penetrating powers is also exhibited by γ-radiation. Whilst the most penetrating can be detected through several inches of lead, the least penetrating can only pass through very thin foils.

Radiation Hazards

It is essential to distinguish between exposure to radiation from an external source, and contamination of the body by radioactive material. External sources only cause damage during the period when the person is near to them, and the rays are attenuated by distance and by absorption in intervening materials. If a person has radioactive material absorbed in his body he is exposed to the unattenuated rays at all times, hence the damage is incomparably greater. 'Protective clothing' is worn, not as a protection against external sources, but to prevent the contamination of the person by radioactive chemicals.

It is sometimes taught that α-radioactive materials are not hazardous to health because of the very short range of the α-rays. This is a very grave error. On the emission of α-rays the nuclei recoil with high energy, scattering the material. Also some α-emitters including radium create radioactive gaseous daughter products. By these processes these materials are easily dispersed and may be absorbed in the body, in which case the relatively heavy α-particles cause grave injury.

There is a widespread belief among science students that ordinary materials will become radioactive on exposure to the rays from radioactive substances. This is not true.

Atomic Structure

This subject involves very difficult physical concepts and a highly complex mathematical theory. It is impossible to treat the mathematics in any satisfactory way below the level of a special honours degree and any reasonably complete analysis requires prolonged post-graduate study. The topic is however so interesting and so important that some form of elementary treatment is demanded. Since the mathematics cannot be discussed at sixth-form level it is obvious that what is required is an introduction to the basic physical concepts.

At the beginning of this century it was shown by the study of gaseous ionization that atoms must consist of electrons combined in some way with a much more massive positively charged structure. A reasonable interpretation of this was that the positively charged part formed a central nucleus about which the electrons moved in planetary orbits. According to classical mechanics these electrons would be continuously accelerated. J. J. Thomson pointed out that, since electromagnetic theory requires that an accelerated charge must radiate, such a system could not be stable. Rutherford and his colleagues by the study of large-angle scattering of α-rays showed that the nuclear idea must be accepted, thereby raising a fundamental theoretical difficulty. In 1913 Bohr published his theory of the gross structure of the spectrum of atomic hydrogen in which he used quantum ideas in an attempt to resolve this problem. During the next twelve years it was fairly generally accepted that an atom consists of a nucleus around which electrons circulated in defined circular or elliptic orbits like planets around the sun. Various attempts were made to calculate energy levels on this basis, with only very limited success. About 1925 the new quantum theory originated and gave immediate and continued success in the mathematical solution of this problem. According to this theory the previous difficulties were the result of using an essentially unsound system of mechanics. The modern view is that the electrons do not have an exactly definable location in space at any instant, and hence cannot be supposed to follow any geometrically describable path. The picture of a regular orbit is not acceptable in any sense, not even as an approximation, but is essentially unsound in principle.

The concept of atomic structure which has now been held for nearly forty years is roughly as follows. There is a central positively charged nucleus around which there are arranged electrons. The arrangements of the electrons cannot be represented in any elementary way, but can be analysed precisely by advanced mathematical methods. There is a stable ground state of the atom from which it may be raised to any of a large number of excited states by absorption of radiation or by collisions. These excited states are usually short-lived and decay to the ground state by the emission of radiation, or occasionally by collisions. The energy levels can be studied in detail by measurements on the emission and absorption spectra and by electron impact experiments. The energy levels can also be computed on the basis of modern quantum theory.

What should we teach about atomic structure in elementary courses? The calculation of energy levels by modern methods is far too advanced to be discussed. The teaching of the historical methods of Bohr and Sommerfeld will not help the students to understand the modern methods at all, and may give wrong ideas of the nature of the problem. What we can do is to show that energy levels are observable properties of atoms, by describing the interpretation of the main features of line spectra together with simple accounts of electron impact experiments.

The Traditional Treatment of Atomic Structure

The idea that atoms contain electrons moving in definable circular and elliptic paths is taught generally as if it were an accepted scientific truth, or at least a first approximation. Students are shown pretty pictures of rosettes which are supposed to represent in some way what an atom would look like if we could see it.

The Bohr theory of the hydrogen atom is commonly taught in mathematical detail. There could possibly be a good purpose in this, for the theory is of great historical importance, but the treatment given rarely fulfills any obvious function. A topic of historical importance can only have meaning if it is set in a historical context. Normally no clear indication is given of the status of this theory, how far it succeeded and how far it failed. Bohr had to make a number of assumptions, some of which were obvious to his contemporaries whilst some were new. Of these assumptions some have been retained to form the basis of later theories, whilst some have been abandoned as unsound. Unless such points are discussed the theory is only misleading and, far from helping the student to understand later ideas, it makes the subject more difficult.

These remarks apply with even greater force to the developments upon Bohr's ideas such as the consideration of elliptic orbits and fine structure. The confused array of *ad hoc* hypotheses which constitutes the old quantum theory is of

absorbing interest to a theoretical physicist who is studying the foundations of his subject, but to the ordinary student it is just a bewildering obstacle to his understanding modern ideas. When teaching planetary motion we do not expect students to conduct detailed calculations on Copernican epicycles before they learn about Kepler's laws. In the same way science teaching must learn to avoid becoming bogged-down in the study of other transitory theories.

The Bohr Theory

The central feature of the Bohr theory is the calculation of the gross structure of the spectrum of atomic hydrogen. The assumptions can be expressed as follows:

(1) The hydrogen atom consists of an electron and a positively charged nucleus of much greater mass.

(2) The electron moves around the nucleus in circular and elliptic orbits exactly definable in terms of ordinary classical mechanics.

(3) Not all the classically permissible orbits are possible, but only certain ones prescribed by quantum rules.

(4) The nucleus and the electron attract electrically, the inverse-square law being applicable at atomic distances.

(5) In the permitted orbits the normal laws of electromagnetic radiation do not apply, that is, an accelerated charge does not radiate.

(6) Those orbits are permitted for which the angular momentum is an integral multiple of Planck's constant divided by 2π.

(7) A transition between two orbits is possible by the emission or absorption of a single quantum of electromagnetic radiation of energy $h\nu$ equal to the difference of energy levels.

On considering this list one wonders what is required by those G.C.E. examiners who ask for a statement of 'the two assumptions of the Bohr theory'.

On the basis of these assumptions, by using the ordinary mechanical equations for circular motion, one obtains a very accurate (but not exact) formula for the gross structure of the spectrum, as is shown in very many textbooks. The assumptions merit far more attention than we can give here, but certain points must be considered.

Assumptions (2) and (4) would be taken for granted in 1913 if (1) were accepted. Assumption (1) was a deduction from recent experimental work, whilst (3), (5), (6) and (7) were new.

The assumptions (1), (3), (4) and (7) are retained essentially unchanged in modern theories, although (4) requires small corrections for very precise calculations. Assumption (6) must be amended. According to quantum mechanics only one component of the pseudo-vector angular momentum can be determined, and the relationship between angular momentum and kinetic energy of rotating systems is different from that in classical mechanics. The quantum number which gives the component of the angular momentum is not the quantum number which appears in the equation for the energy levels as it is in Bohr's theory. Hence, for example, the angular momentum of the orbital motion in the ground state according to Bohr is $h/2\pi$, but according to modern theories it is zero.

Assumptions (4) and (5) are essentially contradictory, as the laws of the electric field are assumed to hold in one respect and not in another. If the electron really orbited around the nucleus then the atom would be a rotating dipole. At any point in the vicinity there would be an electric field which would change periodically, constituting an emitted electromagnetic wave. It is understandable that theoretical physicists such as J. J. Thomson found these ideas unacceptable.

It is of course assumption (2) which is the fundamental weakness of the theory. In the new quantum theory this idea is rejected and consequently assumption (5) becomes unnecessary.

Some Basic Measurements

The most commonly taught method of measuring the electronic charge is the oil-drop experiment of Millikan. This is understandable, since it is unusually direct in character and the principles are easily explained. It is important to appreciate, however, that this method does not have the high status often attributed to it. It is not a very precise method as the errors can probably not be reduced below about a half of one per cent, whilst modern methods have an error of about two parts in a hundred thousand. Further, it was not the first method of moderate precision as is often taught, since it was preceded by the method of comparable accuracy using measurments on black-body radiation analysed by Planck's radiation formula.

The measurement of e/m for cathode rays by J. J. Thomson was a research of great historical importance, although it was very imprecise. The experiments involved the determination of the speeds of the particles, which were found to be typically about a tenth of that of light for the voltages used. It is often taught that this proved that the rays could not be any form of wave motion. This is incorrect, as it only shows that they are not *electromagnetic* waves, and this was already apparent from their deflection in a vacuum by electric and magnetic fields. More recent work has shown that electrons, like all atomic particles, exhibit some wave-like properties.

Chapter 12
MISCELLANEOUS

Elastic Hysteresis

If a wire is loaded in steps to a suitable maximum value and similarly unloaded in steps, it will generally be found that the length for a particular load is slightly greater when decreasing the tension than when increasing it. This elastic hysteresis effect is small for small strains in metals, but large for some other materials particularly rubbers. If a graph is plotted of tension against length the area under the load curve represents the work done in stretching, whilst the area under the unload curve represents the energy which could be given out on relaxing the strain. The area of the hysteresis loop represents the energy dissipated in a cycle.

In a simple experiment to measure Young's modulus for the material of a wire this hysteresis effect is sometimes distinctly observable, although small. The student is not encouraged to notice such phenomena and is explicitly taught to take the average of the extensions on loading and unloading for a particular weight. What is the significance of this procedure? Averages are normally found in two circumstances. The average of a number of separate measurements of the same quantity is taken to obtain a value which is more reliable than any one measurement, and the average of the results of similar measurements on different specimens of the same class is found as when we determine the average height of a population. In the present case the student is taught to average two distinct quantities, namely the length of the wire under one set of conditions and the length under another. This unsound procedure effectively obscures an important phenomenon. It is also pointless, since if it were desired to simplify the subject it would be better to omit the measurements whilst unloading.

Rolling Friction

When one body rolls over another one the state of strain

of both bodies varies continuously. If both bodies were perfectly elastic this would not result in any friction, but real materials exhibit elastic hysteresis which results in energy being dissipated as heat. The resultant friction is quite different in character from that caused by surface effects when two bodies slide or tend to slide over each other, which is normally discussed in physics courses. The almost invariable failure to discuss rolling friction causes students many difficulties, particularly when considering the resistance to motion of vehicles, and problems on the rolling of balls down inclined planes.

The Human Eye

The eye is by far the most important optical instrument, but very little is taught about it in physics courses beyond the correction of defects of vision and irrelevant anatomical terminology. A rough idea of the normal acuity of vision should be regarded as an essential foundation for the teaching of optics, since upon this depends the idea of a clear image.

Certain erroneous ideas about the human eye are widely believed. The retina is often imagined to be a small patch at the back of the eye, whereas it actually covers a large part of the inner surface. Refraction is often supposed to be the effect of the eye 'lens' only, although a large part occurs at the front surface, a fact which has a big effect on vision under water. The enormous range of sensitivity of the eye is rarely appreciated. It is usually taught that the sensitivity is controlled by the pupillary reflex varying the amount of light admitted, although the range of pupil areas is obviously very limited. Actually the sensitivity of the retina may alter by a factor of thousands whilst the pupil radius changes only by a factor of two or three.

The Leaf Electroscope

In its simplest form the electroscope consists of a pair of thin metal leaves mounted on an insulated support. If there is a sufficiently large P.D. between the leaves and surrounding material, such as the floor and walls of the room, the leaves will diverge. Very much greater sensitivity and also freedom from disturbances are obtained by surrounding the leaves by a small earthed case. The leaves and the case then form the two plates of a capacitor, carrying equal and opposite charges. Since the case is so much nearer to the leaves than are the walls of the room the potential gradient is greatly increased. It is this potential gradient which determines the deflection of the leaves, so there is a large increase of sensitivity.

It is thus an essential feature of a satisfactory leaf electroscope that it has an earthed case which carries a charge equal and opposite to that of the leaves. If the case is insulated the sensitivity has the same extremely low value found for leaves without any case. In very many textbooks of physics these essential points are omitted. The case of the electroscope is usually shown uncharged, and the earth connection is often not shown. One reason for this common error may be a misunderstanding of the facts that usually the electroscope has a glass case and stands on a wooden bench. Since glass and wood are normally considered to be electrical insulators it is not obvious that the instrument has in fact an earthed conducting case, but glass and wood are under common conditions of humidity and uncleanliness sufficiently good conductors for the purpose. The question is easily checked by standing the electroscope upon a block of good insulator such as a sulphur or paraffin wax. It is also very instructive for a student to touch the top disc of a charged electroscope with specimens of various common materials which are normally classified as insulators.

Measurement of Low Resistances

The measurement of resistances below an ohm is of great practical importance. Very large numbers of resistors of such values are used as meter shunts, whilst the determination of the resistivities of metals requires measurements of similar magnitudes. It is not generally possible to use such resistors in Wheatstone bridges because of the significant and uncertain errors which result from the resistances of leads and terminal connections. Further, the normal construction of a shunt requires four terminals, two for connection to the main circuit (current terminals) and two for connection to the instrument (potential terminals). The currents drawn through the potential terminals are normally very small, and the resistance between these terminals is well-defined and not significantly affected by leads and contact resistances. The resistance is thus measured between the pair of potential terminals by comparison with the value for a standard four-terminal resistor, using a Kelvin bridge or a potentiometer. The potentiometer method is often taught in elementary work, but it is usually described for two-terminal resistors. The correct method is to pass a current through the two resistors in series, making connections to the current terminals. The potential differences between the pairs of potential terminals are then compared using the potentiometer.

Semiconductors

There are many materials whose electrical conductivity is intermediate in magnitude between that of metals and that of insulators. These semiconductors are of great and increasing technological importance in the construction of rectifiers, transistors and similar devices. Although semiconductors are used in these devices it is incorrect to teach, as is sometimes done, that a semiconductor is by itself a rectifier, or a transistor. One could as well teach that a vacuum is by itself a triode valve.

Perhaps the most generally accepted description of a semiconductor at elementary level is 'a metal which only allows electricity to flow one way'. This is doubly incorrect, since semiconductors are not metals and the description of their electrical properties is completely wrong.

Photoelectric Emission

When a solid or liquid is illuminated with electromagnetic radiation of frequency ν the maximum kinetic energy E of a photoelectron is related to the quantum energy $h\nu$ by Einstein's equation

$$E = h\nu - e\phi$$

where h is Planck's constant, e is the magnitude of the electronic charge and ϕ is the work function of the material.

The product $e\phi$ is the least work required to remove an electron from the material into a vacuum. Inside the material the electrons have a negative potential energy and a range of possible kinetic energies. The electrons which are ejected with maximum speed must be those with the highest kinetic energy within the material.

There are three common errors in the teaching of this subject. One is that E is said to be the energy of any photoelectron, and not just the maximum value. Another is to attribute the emission of electrons of less than the maximum energy to their ejection from deep within the material such that energy is lost on the way to the surface. Finally, it is sometimes taught that ϕ is the potential within the material.

Signs of Increments

In mathematics it is very usual to discuss small increments δx of a quantity x. It is understood that δx represents the change of x, that is if x increases δx is positive, and if x decreases then δx is negative. The new value of the variable is $(x + \delta x)$ in each case. The normal rules of algebra are applied and there is no occasion for any errors of sign. In the teaching of physics unfortunately it is very common to abandon the consistent rules of elementary mathematics, and confusion results.

For example, consider a bulk modulus. According to the ordinary mathematical system this is given by

$$K = \frac{\text{Stress}}{\text{Strain}} = -V \frac{\delta P}{\delta V}$$

The negative sign is required to make the quantity positive, since δP and δV which are the changes in P and V are necessarily of opposite signs. Many physics teachers however say, let δP be the *increase* of pressure and δV the *decrease* of volume. In this case to be consistent the negative sign should be omitted from the definition, since δP and δV now have the same sign. In many cases, however, the negative sign is used, thereby defining bulk modulus to be a negative quantity.

This confusion of sign is particularly serious when there is more than one independent variable. For a triode valve the anode current I_a is a function of the anode potential V_a and the grid potential V_g. The anode slope resistance $R_a = (\delta V_a/\delta I_a)$ with V_g constant, the mutual conductance $g_m = (\delta I_a/\delta V_g)$ with V_a constant, and the amplification factor $\mu = -(\delta V_a/\delta V_g)$ with I_a constant. The negative sign, which is often omitted, is necessary to make a positive quantity, since δV_a and δV_g must have opposite signs to maintain I_a constant. It is easy to show that $\mu = R_a g_m$ by using the theorem of partial differentiation

$$\left(\frac{\delta x}{\delta y}\right)_z \times \left(\frac{\delta y}{\delta z}\right)_x \times \left(\frac{\delta z}{\delta x}\right)_y = -1$$

which is applicable when each of the three variables can be expressed as a function of the other pair considered as independent variables. The same result is often obtained in elementary physics by putting

$$\left(\frac{\delta V_a}{\delta I_a}\right) \times \left(\frac{\delta I_a}{\delta V_g}\right) = \left(\frac{\delta V_a}{\delta V_g}\right)$$

This is incorrect as the cancelling of δI_a is inadmissible. This is because the increments occur under different conditions, different quantities being held constant in each case.

CHAPTER 13

DEFINITIONS

General

There are in physics certain fundamental quantities whose definition presents grave difficulties, and a much larger number of derived quantities for which simple definitions are both possible and necessary.

No useful mathematical or verbal statement can be framed to define clearly and unambiguously what is meant by length or time or temperature, and in such cases any attempt at a general definition is usually unhelpful. It is, however, possible to state clearly what is involved in the measurement of a time interval or of temperature on a particular scale, and a sound understanding of the general concepts can be obtained by a precise consideration of the various important particular cases.

Most physical quantities are related to more fundamental ones in some simple mathematical way and in many cases such derived quantities can be defined in terms of a simple verbal statement or algebraic equation. In this chapter we shall consider one fundamental quantity (mass) whose definition is difficult, and three derived quantities.

Mass

Mass can be defined as the quantity of matter in a body. This leads immediately to difficulties when we proceed to define matter, and it is easy to become involved in complex arguments. This is not necessarily a valid objection to the definition. We are grasping for difficult ideas, and any intelligent discussion may help to clarify the problem. Newton said that mass was the product of density times volume. This definition has been the subject of much undeserved scorn from subsequent generations who assert that as density is usually defined as mass divided by volume this takes us nowhere. It should be clear, however, on the basis of precise experiments such as those of Archimedes which we can perform without having any philosophically sound definition of mass or density that there is a quantity which we may call density which is a characteristic property of any pure material. The mass of a lump of gold is not very definite, different lumps may have different masses and pieces can be added or cut off, but the density of pure gold is a precise constant of nature, even though its value depends very slightly on temperature.

Masses are most simply compared by weighing on a balance, and this point must be stated when explaining the meaning of mass. The danger here is of confusion with some other property of the bodies, such as the gravitational force acting on them. Provided teachers are prepared for such difficulties there should not be much trouble with this point. The distinction between quantity of matter and force is obvious even to a non-scientist. A housewife with no scientific education who was given short 'weight' by a shop would not be satisfied by the argument that it was to her advantage to have less weight to carry.

When given the difficult task of teaching a fundamental idea we must do our best to make it comprehensible by using the concept consistently in intelligible applications. One thing we must certainly not do is to attempt to evade the problem, as many teachers have in the case of mass, by not using the idea. There is an approach to mechanics which avoids the explicit use of mass and instead uses weight divided by g. Apparently it is supposed that the idea of weight is simpler than that of mass. This approach causes great difficulty in the understanding of dynamics. The student may be able to perform correctly the standard types of exercise in which he has been drilled, but he is helpless when confronted with any new application however simple. In any rational system of mechanics g will only appear in problems involving gravitation, but in this sytem it appears in problems which do not involve gravitation.

Students can then hardly be blamed for multiplying or dividing by g any result which does not look right. The gravitational acceleration has hence acquired a mystic significance which is well illustrated by the results given by a group of university engineering students who were asked to express the British thermal unit in gm calories. There were 23 different answers ranging from 209,000,000 to 0·0008117, mostly obtained by various operations involving g.

Specific Heat

If a quantity of heat Q is supplied to a mass M of a substance thereby raising its temperature from T_1 to T_2 we define the mean specific heat $\overline{C}$ of the substance over this range of temperature by the equation

$$\overline{C} = \frac{Q}{M(T_2 - T_1)}$$

The value of $\overline{C}$ depends upon the conditions (*e.g.* constant pressure or constant volume) and these must be specified.

The value of $\overline{C}$ generally varies with temperature, so to define it at a particular temperature, we consider an infinitesimal heat increment δQ and put

$$C = \frac{1}{M} \frac{\delta Q}{\delta T} \qquad \ldots \text{(13.1)}$$

Elementary courses provide a remarkable variety of definitions of specific heat. Perhaps the most usual is 'the number of calories required to raise the temperature of one gram of the material by one degree'. In objection to this it may be observed that the concept of specific heat does not depend upon the calorie or any other particular unit, and also that as C sometimes changes dramatically in value in a range of one degree such a finite temperature interval should not be included in the definition. Another approach is to state equation (13.1) in some form and call C a constant, which generally it is not.

Sometimes specific heat is defined as the ratio of the thermal capacity of the material to that of an equal mass of water. This ratio has value in certain types of work but its importance in most branches of physics has largely vanished.

Its use sometimes causes confusion in the dimensions of thermal quantities.

Resistivity

When a current flows parallel to the axis through a cylinder of material of length L and cross-sectional area A the resistance R is given by the equation

$$R = \rho \frac{L}{A} = \frac{L}{\sigma A}$$

where ρ is called the *resistivity* or *specific resistance* of the material, and its reciprocal σ is called the *conductivity*. Clearly the dimensions of ρ are those of resistance times length, the M.K.S. unit is the ohm metre.

A very common 'definition' is that the resistivity of a material is 'the resistance of a unit cube'. This is a typical example of a meaningless statement taught as a definition. Nobody ever constructs cubic blocks of unit size in order to measure their resistances, so the 'definition' is unrelated to a real situation. Experiments are performed on specimens with non-unit dimensions and the 'unit-cube' concept tells us nothing about resistivity in such a case.

Consider for example the function F defined by

$$F = R^{A/L^2} \exp\left(\frac{A - L^2}{17A}\right)$$

according to the unit cube definition this function is the resistivity since it is equal to R for a unit cube! There is of course a very large number of distinct functions of this sort, all equal to R in this particular case.

Not only is the unit cube idea meaningless, but it is definitely misleading. Many students understandably confuse resistance per unit cube with resistance per unit volume, which leads to mistakes which nobody could conceivably make if they were not taught in this way. Teachers and authors have to introduce carefully devised examples to prevent this type of confusion, which would not be necessary if the correct definition (and no other) were taught in the first place.

The unit cube approach leads to difficulties with units, especially when it is necessary to convert from one unit of length to another.

Equilibrium Vapour Pressure

It is generally known that all liquids tend to evaporate, but it is far less commonly known that solids do so also. The sequence solid-liquid-gas is taught so generally that many students are unaware that solids can evaporate directly, and some are quite convinced of the impossibility of this. There are several commonplace examples of the evaporation of solids and of condensation directly to the solid phase which should be taught to all science students. These include the formation of hoar frost, the evaporation of ice in strong winds and 'freeze drying', the sublimation of iodine and of tungsten lamp filaments, and the sublimation of solid carbon dioxide, and many organic solids such as naphthalene.

Both liquids and solids can exist in equilibrium with their vapours, the vapour molecules leaving the condensed phase at the same rate as they return to it. In these circumstances the pressure exerted by the vapour is called the *equilibrium vapour pressure* (E.V.P.). An alternative name is *saturation vapour pressure* (S.V.P.). This is rather less suitable, since it conveys less meaning than E.V.P. Another term sometimes used, *maximum vapour pressure,* is very misleading since a vapour at a particular temperature can exert more than its equilibrium pressure. For a vapour to condense at a finite rate it is clearly necessary for the pressure to exceed this 'maximum' value.

Equilibrium vapour pressure is usually given most unsatisfactory definitions. Very commonly it is said to be 'the pressure exerted by a vapour in contact with the liquid' (solids are hardly ever mentioned). Sometimes this is slightly improved by adding 'in a confined space'. The absurdity of such definitions should be obvious on considering the conditions in a room in which there is a vase of water. Obviously it is most unusual for the space in the room to be saturated, and the water will normally remain for many hours evaporating slowly. The room as a whole may have humidity far below 100 per cent and even in the immediate vicinity of the water surface the vapour pressure must be below the E.V.P. or the liquid would not continue to evaporate. Even in a very small confined space the vapour pressure will not generally have the equilibrium value. If the temperature is rising the vapour pressure will be below the E.V.P. and *vice versa*. If there is a steady state with a temperature gradient the volatile material will tend to distil from the warmer to the cooler parts. In this case there is obviously no question of equilibrium, although if the distillation is very slow the vapour pressure becomes almost exactly equal to the E.V.P. at the temperature of the coldest part of the vessel.

It should be clear that the vapour in contact with a condensed material can only be guaranteed to have the E.V.P. if the conditions are those of thermodynamic equilibrium.

According to the traditional method of teaching physics a student is told that a vapour in contact with a liquid exerts a particular pressure at a given temperature. He is then given descriptions of evaporation and condensation processes in which it is necessary to assume that the pressure exerted is less or greater than this value. This is a typical example of the lack of consistency which permeates physics education.

Discussion

The examples of definitions discussed in the three previous sections should be considered together with those other cases which have been treated elsewhere in this book, particularly thermal conductivity, entropy and electrical resistance. These examples illustrate the grave failure of traditional physics teaching to give clear and meaningful definitions. It can be shown that nearly every so-called definition given in most school physics textbooks is either incorrect, ambiguous or just meaningless. Most physical quantities can be defined in the same way as specific heat, resistance and resistivity, by a simple algebraic equation combined with a verbal statement of the meaning of the terms and the conditions assumed. It is generally both unnecessary and undesirable to attempt to express such algebraic formulae in words, and reference to units is irrelevant in a definition. It is disturbing to find that when a quantity is defined by a mathematical formula in a textbook and a verbal interpretation is then given, the verbal statement very rarely means the same as the equation. Apparently large numbers of honours graduates in science believe that the formula $x=yz$ means just that 'when the quantities y and z are unity then x is equal to one'. It is also widely believed that such an equation means that x is proportional to z, even though y is not constant.

The proper purpose of a definition is to explain the meaning of some quantity or idea and to enable the student to recognize it in any reasonable circumstances. This purpose is obviously not achieved by stating the value of a function in some highly improbable particular circumstances such as for a unit cube.

The common failure to give correct definitions for such functions as equilibrium vapour pressure and entropy appears to reflect a lack of understanding of the relevant phenomena, and of the use of these functions in the description and analysis of them.

CHAPTER 14

THE UNREALITY OF G.C.E. PHYSICS

Physics as an Algebraic Exercise

An investigation of G.C.E. examination questions and of the numerical examples in many textbooks suggests that physics is widely regarded, not as a study of the physical world, but as a source of problems in elementary algebra.

We have seen in Chapter 6 that the essential physical principles of expansion processes are generally ignored at this level and such formulae as $PV^\gamma =$ constant are used as a matter of course in situations to which they do not apply. Many questions demanding the ignorant use or misuse of these formulae are set in G.C.E. and similar examinations, and in consequence those students who later study more advanced thermodynamics meet great difficulties. What is the purpose of such absurdities in physics examinations? Why should a student be expected to memorize equations the applications of which are unknown to him, his teachers and his examiners? And what place has second-form algebra in an A-level or S-level physics examination, unless a prerequisite of its use is an understanding of the physical principles of the subject?

Similar remarks apply to the very large numbers of absurd exercises on thermal conduction. Very many of these questions involve heat transfer through sheets of solid material between fluids as discussed in Chapter 7. In such cases the results obtained are often one or two orders of magnitude higher than would be found in practice, and sometimes even more. There is a recently published example of a problem concerning a domestic hot-water tank which gives the result that the tank loses heat to the surroundings at a rate of *sixteen million watts*. This is achieved without any temperature difference between the outer surface of the tank and the surrounding air! Another type of heat conduction problem is exemplified by an A-level question on the conduction of heat along the nickel wires supporting a lamp filament. The ends of the wires were assumed to be at the temperatures of 2200°C and 150°C respectively and the temperature gradient had to be assumed to be uniform. The unreality of the problem is shown by the fact that nickel melts at about 1450°C. On considering a more refractory wire with its ends maintained at these temperatures it is easily found that nearly all the heat conducted into the hot end is radiated from the sides of the wire, and hence the idea of a uniform gradient is absurd.

In a recent O-level examination an ammeter reading to 3 amps was said to have a resistance of 700 ohms. The question required the calculation of the shunt required to read to 10 amps, and it was also asked why the instrument was not suitable for use as a voltmeter. Simple arithmetic shows that this meter would work at 2,100 volts and dissipate 6·3 kilowatts without the shunt. It would clearly not be suitable for *any* purpose.

Another type of algebraic problem which ignores physical reality arises from the use of the lens formulae. For example, a question may be set about a camera which is said to be in focus for certain distances of the object and film from the lens. The object distance is changed and it is asked through what distance the film must be moved in order that there shall again be a sharp image. The underlying assumption is that there is only one exact point at which an optical instrument will give a sharp image of a particular object. The falsity of this assumption should be obvious to anyone who has ever tried any simple optical experiments, and in particular to anyone who has ever used a box camera. The teaching of geometrical optics has become dominated by pointless calculations of object and image distances, whilst the vitally important elementary physical facts of the subject such as acuity of vision and depth of focus are ignored.

The custom of using physical formulae as a basis for algebraic exercises without taking account of physical reality is found throughout elementary physics. One of the less serious, but particularly ridiculous, manifestations of this is the practice of devising problems and experiments which reverse the normal order of things. For example, in learning mechanics a student might usefully measure the periodic times of pendula of various lengths, amplitudes and materials. Instead of this he is required by the algebraic heresy to assume a formula and hence find the height of an inaccessible object to which a pendulum has been attached by magic art.

Applicability of Laws

It is essential to appreciate that our complaint against the traditional treatments of problems such as those on expansion of gases and heat flow is not that they are inexact, but that they are unsound and unrealistic. In all quantitative work it is necessary to make approximations, and it is particularly important in elementary work that a student should not have his attention distracted from the principles of the subject by excessive concern with minor corrections. In these problems however the aspect which receives most attention is often that which is of least importance. For example, in the passage of heat through a window the temperature drop in the glass which is traditionally the only factor to be considered is actually a small correction, whilst the temperature drop in the surface air which is normally ignored is almost the whole effect.

It is of the utmost importance that it should be made absolutely clear in teaching whether a particular law is exact or approximate and whether it is universal or of only limited application. Over a period of many years large numbers of students have come to study with the author after achieving high success in the sixth-form. They have all known the formal statements of very many physical laws, but very rarely has any one of them shown any knowledge of the accuracy and applicability of these laws. Thus many students believe that momentum is not conserved in inelastic collisions, that Ohm's law is exactly true for all substances and that diatomic gases have gamma equal to exactly 1·4. In most cases however when a student is asked whether or not a law is exact, and whether or not it is universally applicable he can only guess wildly. The idea that such a question could arise is obviously quite new to most students to whom physical laws are just slogans to be memorized for an examination.

The degeneration of physics teaching in recent years is

illustrated by the growing custom of teaching the laws of ideal gases as the laws of real gases, without ever distinguishing between the real and the ideal, and without ever using the terms 'ideal' or 'perfect'.

The Electromagnetic Doppler Effect

To illustrate the importance of knowing whether a law is exact or approximate we will consider the Doppler effect for electromagnetic radiation. In the opinion of the author this topic is too advanced for schools, but it is frequently taught and very many numerical problems are set upon it, so it merits our attention.

This effect is a change in the observed wavelength of an electromagnetic radiation resulting from relative motion between the observer and the source. Let the velocity of the source as seen by an observer O be V, with a component V_R along the line joining the source to the observer. V_R will be taken as positive if the separation is increasing. Then the wavelength as determined by O will be λ given by

$$\lambda = \frac{\lambda_0 (1 + V_R/c)}{\sqrt{1 - V^2/c^2}} \qquad \ldots \ (14.1)$$

where λ_0 is the wavelength determined by an observer at rest with respect to the source, and c is the speed of electromagnetic waves in vacuum.

Equation (14.1) expresses the exact law given by the special theory of relativity. Provided V is much less than c and not very much greater than V_R the law can be written in the approximate form

$$\lambda = \lambda_0 (1 + V_R/c) \qquad \ldots \ (14.2)$$

Equation (14.2) is commonly taught as if it were the exact law. This may be partly because the analogous effect in acoustics is represented by this equation in a particular case, and it is assumed that the result must also hold for electromagnetic waves, although the principles involved are very different.

Provided the conditions stated above are satisfied we can use the approximate equation confidently. This will apply to the use of radar to measure the speeds of vehicles and missiles, since these speeds are known to be relatively low.

It is common however to set problems on astronomical motion where such assumptions are not necessarily valid. Usually the question gives the difference of wavelength observed between a spectral line obtained with light from a star or nebula and the same line produced by a terrestrial source, and require the calculation of the velocity of the star or nebula. Even if equation (14.2) is valid it is only possible to find V_R, that is one component of the velocity. The examiners usually forget that velocity is a vector quantity. Further, unless other information is given, the problem is insoluble since there is an infinity of pairs of values of V and V_R to give any particular ratio of λ/λ_0 in equation (14.1). As an example consider a scholarship question in which the red-shift of light from a nebula was given and the candidates were asked to calculate the velocity and show that the nebula was moving away from the earth. This is the only possible conclusion from equation (14.2) but the exact equation (14.1) shows that a large transverse component of motion will give a red-shift without V_R being necessarily positive. Here is a case where the results given by an exact equation may even be qualitatively different from that given by the approximation.

It was stated above that one condition for the use of the approximate equation is that V shall not be very much greater than V_R. A few years ago the results of a complicated and much-publicised research to measure the gravitational red-shift were invalidated by the neglect of this condition. The research involved the very precise measurement of the wavelength of gamma-rays emitted by iron nuclei in a solid. The emission of one of these quanta takes place over a time (10^{-7} secs.) which is very large compared with the period of oscillation of the ions in the crystal, hence the value of V_R for a stationary observer averaged over the period of emission is zero. For this reason the Doppler effect resulting from thermal oscillations was assumed to be zero. But the average value of V^2 is finite, hence an important temperature-dependent effect does in fact occur.

Invented Results

A particularly serious consequence of the traditional neglect of physical principles is the invention of imaginary experimental results which are taught as observed facts. In practice a standard formula is used in a field where it does not apply and a result is calculated. It is then taught that this result has been established by experiment. Ultimately these supposed experimental results may be used to support the original unsound theory.

As was shown in Chapter 2 an unsound assumption concerning the density of the earth leads to the erroneous conclusion that g decreases on going down a mine. In consequence it is now sometimes taught that this is an experimentally observed fact.

The invalid application of the law of Wiedemann-Franz-Lorenz to non-metals leads to the idea that they must all be thermal insulators. Hence it is often taught that experiment shows that all good conductors of heat are also good conductors of electricity.

By using geometrical optics it is deduced that the image produced by a pin-hole becomes sharper without limit when the hole is made smaller. This if often taught to be the observed effect, although wave theory and experiment give a different result.

The energy levels of the hydrogen atom can be calculated easily from a simple formula. On this basis various accounts have been written of experiments in which hydrogen gas is excited and ionized by electrons. The inventors of these accounts have overlooked the fact that under ordinary laboratory conditions hydrogen consists of diatomic molecules which have quite different energy levels.

Similar accounts have been written of excitation and ionization of monatomic substances such as metal vapours. Such accounts often ignore the fact that electrons from hot filaments have a range of initial energies and that their acceleration between electrodes is affected by contact potential differences. Hence the described results are very different from those really obtained.

Materials

In the world of traditional school physics the properties of materials are those observed in the temperature range from 0°C to 100°C with a pressure of exactly one standard atmosphere. In some examinations when boiling water or steam are mentioned the student is expected to know that the temperature is 100°C. Actually water will boil at almost any temperature within the range in which it is liquid, that is from near 0°C to about 380°C. Of course there is no definable upper limit to the temperature of steam.

It is taught that hydrogen and oxygen are gases, mercury is a liquid and sodium and tungsten are solids. We have seen (Chapter 7) that many writers assert that mercury is the only liquid which is a good conductor of heat. Presumably these writers must be aware that all metals exist as liquids, but the fact has not really entered into their thinking, and according to the traditional outlook mercury is the only liquid metal. Students are worried on meeting a reference to the gas constant for tungsten. How, they ask, can it have a gas constant when it is a solid?

It is taught almost universally that a gram-molecule of any gas at S.T.P. occupies 22·4 litres. This is of course absurd as most substances do not exist as gases under these conditions, and those that do so do not obey the gas laws exactly. The true rule is that if the density of a gas is measured at a low pressure and the result is corrected to S.T.P. using the ideal gas equation, then the corrected volume per gram-molecule will have this value. This statement is harder to learn than the usual one, but it has the merits of being true and meaningful.

Another aspect of the narrow view of the properties of materials is exemplified by the very commonly taught idea that the viscosity of all fluids decreases with increasing temperature. Actually the viscosity of a gas increases with increasing temperature as does that of certain liquids, namely liquid helium and (over a certain range) sulphur.

Tables of values of physical properties of matter are often given in textbooks without any indication of the conditions under which the values are applicable. Temperature is nearly always important and should be stated, as should any other relevant factor such as for example the frequency of the A.C. used to measure a 'dielectric constant'.

Finally it should be noted that *glass* is not a particular

substance but a *type* of substance. There are many glasses, each with its own characteristic physical properties.

Obvious Errors

Even the limited experience of the schoolboy should enable him to challenge some of the unreal ideas of his teachers and examiners.

If he has ever watched raindrops on a window he should have noticed hysteresis of the angle of contact and will therefore know that this is not a 'definite angle' as it is described in textbooks.

If he has normal colour vision and has examined the spectrum of white light he will have failed to distinguish any distinct colour 'indigo' between blue and violet.

If he has normal vision he will notice that he can see objects clearly at distances considerably less than 25 cm. which he is taught is the least distance of distinct vision.

If he owns a portable radio or rides a bicycle at night he is accustomed to employ Leclanche cells for long periods of continuous use. His textbooks tell him that such cells are not suitable for continuous operation, and he is expected to explain why this is the case in examinations, but he still uses them.

CHAPTER 15

ANCIENT AND MODERN

Advancing Science

The normal process for bringing a physics syllabus up to date is the addition of new material to its end. Since the length of study and the capacity of brains do not increase significantly this generally means either that the new topics are not taught, or that the time available for the older topics is reduced. The latter policy usually involves a gradual decrease in the thoroughness with which the essential foundations of the subject are taught.

New discoveries are made by post-graduate research workers, who when they publish naturally assume in the reader a full body of graduate knowledge such as they themselves were taught as students. If new ideas are to be incorporated in undergraduate and more elementary courses it is necessary to find some way of teaching them which does not require the assumption of all the older material for which there is no longer sufficient time. Obviously from time to time a radical reform of the teaching of the subject must be necessary, with redundant material being removed from syllabuses and the teaching of fundamentals being replanned to provide a satisfactory foundation for the understanding of new ideas.

Keeping up to date in this way is exceptionally difficult in universities, but it is a problem which does not arise to the same extent in more elementary courses where the topics taught should be mostly basic principles of lasting validity. The fantastically confused teaching of Newtonian dynamics which is so common can hardly be attributed to teachers not being up to date. There are, however, some topics of which the traditional treatment is marred by excessive conservatism. The case of atomic structure has been considered in Chapter 11 and certain other examples are discussed below.

Calorimetry

By 1850 it had been established that heat is a form of energy. By 1900 it was possible to determine quantities of heat absolutely by electrical measurements more accurately than comparative measurements could be made by the method of mixtures. The electrical heating method has since then been generally the most straightforward and accurate method of measuring most thermal quantities. Despite this, in the nineteen-sixties, students are still taught other techniques of calorimetry in far more detail than the electrical heating methods.

Thermal measurements are now made commonly both at high and at very low temperatures. The nineteenth century idea of relating thermal quantities to the specific heat of liquid water is thus pointless for the majority of applications. Although biologists may still have some reason for using calories it is far more sensible for physicists and chemists to measure heat in joules and heat flow in watts, as is slowly becoming general in advanced work. The adoption of these units for the majority of elementary calculations would be a simplification, would help students to appreciate that heat is a form of energy (which is little understood at present) and would eliminate the conversion factor J from equations. It would also prevent the mistakes which arise from confusion between gram and kilogram calories. In connection with thermal units it may be noted that some very strange customs have arisen. For example, when heat is radiated the C.G.S. unit is correctly said to be the erg, but when it is conducted the C.G.S. unit is said to be the calorie. Much confusion would be avoided in this and other fields if units were always correctly identified and quantities were never expressed as so many 'C.G.S. units'.

It is very rarely appreciated that the experiments often called measurements of the mechanical equivalent of heat are absolute measurements of the specific heat of water.

Photometry

In elementary courses this subject often comprises a catalogue of primitive types of photometer heads and of obsolete flame standards. It may be noted that the incandescent filament lamp standards were introduced in 1909 whilst in 1948 there was adopted a new unit of luminous intensity, the candela, based on the light emitted from a black body source at a specified temperature.

D.C. Electrical Measurements

The system of electrical units and measurements has for many years been based on the absolute determination of the ampere, using a current balance, and of the ohm, using the Lorenz method. Portable standards are calibrated resistors and the Weston cadmium cell whose E.M.F. is determined by means of the current balance used in conjunction with a resistance standard.

Given these portable standards and a precision potentiometer an ordinary laboratory can conduct a whole range of measurements of current, voltage and resistance, and the calibration of ammeters and voltmeters.

The traditional school electricity course belongs to another age. The old unit of current based on the electrochemical equivalent of silver, and that of resistance based on the resistivity of mercury are still encountered. The laboratory contains tangent galvanometers, metre bridges and potentiometers consisting of a wire stretched on a metre rule. The Daniell cell is still sometimes used as a standard of E.M.F. No doubt shortage of money may partly account for the use of archaic equipment, but wire bridges are often fairly expensive and always make excessive demands on space on benches and in cupboards. Moderate quality resistance boxes and standard cells are not very expensive and take up little space, and their use is far more instructive than that of traditional equipment.

The author has found that university freshmen can rarely connect correctly the simplest circuits even when given a diagram. This is easily understood when one considers the type of equipment with which they have usually been trained. The usual wire bridge is so designed that the various components can be joined and the prescribed measurements performed without the student having necessarily understood the circuit in the least. To understand the Wheatstone bridge it is best to construct such circuits using a resistance box and a few resistors of known fixed value to measure various widely different unknown resistances.

Similarly students who are taught using the usual wire potentiometers seldom acquire much understanding of this most important circuit. It is more instructive to construct a potentiometer with two resistance boxes in series. Such methods teach the principles of the circuits far more clearly than does the use of wire bridges which often seem designed to obscure the electrical considerations. Further the use of resistance boxes permits high precision and thereby teaches the *purpose* of null methods.

It should be noted that it is incorrect to compare the E.M.F. of a cell with that of the potentiometer driving cell, as is often done.

The old-fashioned approach to the Wheatstone bridge usually leads students to believe that 10,000 ohms is a very high resistance and about the limit which the circuit can measure. Actually using modern null-detectors, resistances of up to 10^{12} ohms can be determined this way.

Atomic Physics

The subjects of atomic and nuclear physics are commonly taught under the most inappropriate title 'Modern Physics'. There is in fact very little which is modern about most courses in these subjects below the level of a special degree, as most of the information which is taught refers to very early work. This is particularly unfortunate since so much of these subjects is very much easier to understand when seen from a modern standpoint. For example, it is easy to understand the ionization of molecules when electrons are projected into a gas at very low pressure as in the ion source of a modern mass-spectrometer, but it is extremely difficult to approach this topic as the pioneers had to do by interpreting the complex processes of gaseous discharges at relatively high pressure. Many teachers begin the teaching of atomic physics by a detailed account of the phenomena of the discharge tube. The purpose of this is obscure, since the observation of various glows and dark spaces is of very little help in understanding ionization processes. In this connection it may be noted that often no explanation is given for the emission of electrons from the cathode in a discharge tube.

There are scores of methods of measuring e/m for beams of charged particles, several of which are commonly taught at scholarship and general degree levels. With such a large choice of methods it is sad to find that the oldest receive most attention. Of course, all students should learn about the discovery of the electron by J. J. Thomson because of its great historical importance, but when it is decided to teach a method for the variation of mass with speed it is surely not necessary to choose the method of Bucherer (1909). In 1938 Zahn and Spees performed a more precise variation of this method which is simpler to understand than the original as well as having the further interest of being also applied to the positron.

Any account of mass spectra should refer to the pioneer mass spectrograph of Aston (1920) but it is only desirable to teach those features which are of real historical importance. Far too often the student is burdened with mathematical details of the method of focusing. As this method is very different from those used in modern instruments such details are not very instructive. Other instruments described in elementary textbooks are usually those constructed before 1930, and the only applications described are the measurements of the masses and relative abundances of isotopes. Except in very advanced books it is rare to find any mention of modern instruments such as the mass spectrometers pioneered by Nier with their wide range of applications in physics and chemistry, and yet these instruments are no more difficult to understand than the older ones.

One application of the modern mass spectrometer is the study of processes of ionization of molecules by electron impact. This is a field in which much important work has been done in recent times but it is rare to find in any textbooks references to experiments less than forty years old.

Gamma for Gases

The ratio C_p/C_v, usually denoted by γ, has a value very near to 5/3 for monatomic gases at all temperatures ordinarily attainable. For all other gases the value is normally lower than this, and decreases slowly with rising temperature. For hydrogen γ is about 5/3 up to about 60°K. It falls to about 1·4 at room temperatures and decreases further at higher temperatures. Several other diatomic gases have gamma about 1·4 at and below room temperatures, the value decreasing slowly at higher temperatures. Gases with more complex molecules have lower values about which the only general rules are that they decrease as temperature rises and that the more complex and less strongly bound molecules give the lowest values.

These results are fully explained by quantum statistical mechanics in terms of the equilibrium probabilities of occupation of rotational and vibrational energy levels. Before the introduction of the first quantum ideas in 1900 attempts had been made to develop a system of statistical mechanics in accordance with classical mechanics. This theory which involved the idea of equipartition of energy between various degrees of freedom proved powerless to predict even approximately the values of specific heats except by the assumption of highly improbable hypotheses. The results of this theory are still, however, regarded by many physicists as firmly established truths. A typical textbook account of the specific heats of gases may be expressed as follows, 'The values of gamma for monatomic, diatomic and triatomic gases are 5/3, 7/5 and 9/7 respectively. If gamma does not have one of these values it proves that some of the molecules have different numbers of atoms than those normally present'. The values quoted are those given, on certain assumptions, by the equipartition theory. Presumably those teachers who believe in these results are unaware that γ is highest at low temperatures, or they would have to conclude (as some students do) that gases are dissociated by cooling!

It is true that many diatomic gases have values of roughly 1·4 and it is usual to teach that this is the figure given by equipartition theory. This assumes that the molecule has three degrees of freedom for translational motion and two for rotation *and no others*. It requires that the molecule cannot vibrate, although it consists of two particles with finite mass bound by an elastic interaction. The classical theory cannot account for the absence of a vibrational effect in such a case, nor for the absence of any effects from the multitude of other forms of molecular energy revealed by spectroscopy.

It would be better to teach students the simple empirical facts in this case and to refrain from theoretical interpretation, rather than (as is now done) to teach an ancient and improbable theory with certain distortions of the facts to improve the agreement.

Similar remarks apply to other applications of statistical mechanics which are very often taught on an extremely outdated basis, for example the theory of the specific heats of solids.

Chapter 16

SUMMARY AND CONCLUSIONS

Errors of Fact

The number of incorrect 'facts' commonly taught in elementary physics is very large, being much greater than can be discussed in this book, which is more concerned with general principles than with individual errors. This applies particularly to atomic physics, where different authors and teachers cover more varied material than is usual in other fields, and to geometrical optics, where far too many diagrams would be needed for satisfactory discussion. Now, everybody makes many mistakes and it is to be expected that every student will receive some misinformation. In every subject students must learn that no teacher, lecturer, textbook or examiner is infallible, and that some few of the accepted truths of today may be the exploded fallacies of tomorrow. What then are the objections to the factual mistakes in physics teaching?

The trouble is that far too many mistaken ideas are generally accepted as scientific truths. When a mistake occurs in a textbook or in an examination paper this should be noticed and corrected, instead of which such errors are commonly copied and perpetuated. The phrases 'it is in all the books' and 'it is always taught that way' are generally considered to be irrefutable arguments.

The author has often been told that the mistaken ideas taught in elementary courses do not matter as they will be corrected in later teaching. This hope is no doubt generally derived from a misunderstanding of the nature of the errors. According to what is sometimes called the 'progressive' method of science teaching a topic is treated successively at increasingly complex levels in different stages of a course. At the first stage a principle may be applied to the simplest cases, minor disturbing influences being ignored. In later stages more difficult applications will be considered with details of corrections. Such a procedure is not only correct, but it is often essential to work this way. There is nothing wrong with a first treatment which considers only simple cases, provided the relevant facts and the principles are stated correctly from the beginning. There is all the difference in the world between a valid approximation and a mistake.

This question is of great importance, and as it is often misunderstood it may be helpful to consider an analogy. Suppose that we establish radio communication with the inhabitants of another planet and teach them our language. These very different creatures may be surprised to learn that we walk upright and that we are of two sexes, and they may ask about the relative heights of men and women. We could send them various alternative replies.

(a) A complete set of statistics on the heights of men and women of all ages and races with details of the effects of diet, environment and disease.
(b) 'On the average men are six per cent taller than women, but heights vary and a tall woman may be taller than a short man'.
(c) 'Men are taller than women'.
(d) 'All men are taller than all women'.
(e) 'All women are taller than all men'.

Such a reply as (a) would be of little use and a sound 'progressive' teacher would send reply (b) in the first instance, giving further details later if required. Reply (c) would not be satisfactory as it is too vague. No rational person would argue that as complete information could not be understood straight away we should send such a reply as (d) or (e), yet this is the justification which is given for the teaching of similar errors in physics.

We must further question the hope that later teaching will correct the wrong ideas which are taught early on. Usually the errors are not corrected even if the students do study physics further. The mistakes of elementary physics are not just the ideas given to sixth-form pupils. They are the ideas which are still believed by the honours graduates who teach them in all good faith. This book presents fundamentally not a condemnation of our schools but of our universities.

Errors of Reason

Far more important than the particular facts, true or false, which are taught to schoolchildren and older students is the intellectual training received. Educationalists assert that we must teach young people 'how to think'. The traditional teaching of elementary physics must be examined from this point of view.

In practice there is often a complete failure to provide a rational treatment of the subject. In dynamics Newton's laws are stated and then implicitly contradicted. The law of gravitation is commonly stated in terms which lead immediately to nonsensical predictions. The ideas which are almost universally taught about intermolecular forces are obviously inconsistent with the continued existence of matter. The mass-energy relationship is often said to apply to all kinds of energy and is then used to distinguish nuclear energy from other kinds. In gases molecules are assumed to exert a pressure without exerting any forces. The principles of thermometry as usually taught are a farrago of nonsense. When heat flows from a solid to a fluid, sometimes it is assumed that there is a temperature drop and sometimes it is assumed that there is not, without any discussion or explanation. Certain laws are deduced for all 'adiabatic' expansions of gases, then quite different laws are assumed in particular cases. Attempts to define important quantities result in ambiguous or meaningless expressions. Definitions and physical laws are often confused.

It is commonplace to find statements in textbooks ignored or contradicted in later chapters or even in the same paragraphs. On reviewing sixth-form physics the dominant impression is that of confusion of thought. What is the effect on the minds of young people who are taught *to think wrongly* in this way? If a student is regularly taught sound simple logical arguments with clear definitions and the consistent application of principles there is no certainty that he will become a brilliant thinker, but it is to be expected that his powers of reasoning will develop. If instead he is regularly taught to regard absurdities as truths and contradictions as reason, then grave harm must result.

Criticism

There are two popular fallacies about criticism, one is that it is easy to criticize, and the other that criticism should be constructive.

If it were easy to criticize far more people would do it. As it is, the sad state of physics education is clear evidence of an almost total failure of criticism. Many absurd questions are set in examinations. According to the examiners whom the author has consulted about some of the worst cases, not a single complaint has been received from any of the scores of teachers whose pupils are affected. Serious errors which must gravely mislead students are published in textbooks and remain uncorrected for ten or twenty years or more whilst the books pass through several editions. Any successful sixth-form textbook must be read by hundreds of teachers each of whom has the moral obligation to point out errors. Either they lack the knowledge, or the will, to criticize.

The type of criticism which is most needed in our present circumstances is destructive, since such constructive criticism as there has been has proved of little avail. Nearly every topic which is discussed in this book is treated correctly in some one or other of the many elementary textbooks, but these correct treatments do not receive any general acceptance. It is obviously not sufficient to state that something is true, instead we must state quite unambiguously that something else is false.

Current Reforms

It has been obvious for many years that there is much that is wrong with physics education, and there have been various very limited attempts at reform. To illustrate the unsatisfactory results which have been achieved we may consider one feature of a new O-level syllabus which was introduced a few years ago. Nuclear energy was included, apparently in order to be up to date, and hence the mass energy relationship had to be taught since this was thought to be necessary to explain this subject. This is of course an extremely difficult concept, especially to pupils at O-level who have usually very little understanding of either mass or energy. The new syllabus made a bad situation worse by excluding from the course the simple calculations in mechanics which are so necessary to acquire any understanding of the idea of energy.

More recently a more general reform has been undertaken. This is to be welcomed, but it is to be noted that this has so far been directed essentially towards a change of teaching methods rather than of the ideas which are taught.

One very disturbing feature of the new approach to physics teaching is the emphasis which is placed upon pupils 'finding things out for themselves' instead of being taught dogmatically. Of course all good teachers have always encouraged their pupils to make discoveries themselves wherever this is possible, but in physics this technique is necessarily of very limited application. Physics is a science which has been developed over centuries by the life work of thousands of men and women, many of them being outstanding geniuses. There is very little in this immense structure of knowledge which can be taught other than dogmatically. No child can look at a planet through a telescope and deduce the laws of motion, or operate a toy steam engine and discover the laws of thermodynamics. Difficult scientific ideas cannot be deduced by childish reasoning from simply observable phenomena.

An example of the non-dogmatic approach is the use of a demonstration of the Brownian motion to establish the atomicity of matter. In fact, all that a qualitative study of this phenomenon can show is that matter which is macroscopically at rest is microscopically in turbulent motion. The evidence for the existence of atoms is far too profound for young schoolchildren to understand.

Conclusion

A radical reform of physics education is necessary. This cannot just be a change of syllabuses and of teaching methods. The task is the re-education of a profession.

APPENDIX

Notes and References

Chapter 3. For a discussion of relativity dynamics and the derivation of equation (3.1) see for example W. H. McCrea, *Relativity Physics,* Methuen.

Chapter 4. The interactions of molecules are discussed in detail by E. A. Moelwyn-Hughes. *States of Matter,* Oliver and Boyd. This work also treats the relationship between the equation of Van der Waals and the law of force.

Chapter 7. Convective heat transfer is generally treated in much greater detail in textbooks of mechanical engineering than in textbooks of physics. See for example M. Fishenden and O. A. Saunders, *An Introduction to Heat Transfer,* Oxford University Press.

Chapter 11. A sixth-form textbook which correctly describes positron emission and gives an account of the elementary systematic rules of radioactivity has been published recently by Dent. This is Part 4 of *Concise Advanced Physics* by D. R. J. Wood.

The types of stable and unstable nuclides and their main properties are very conveniently summarized in the form of a chart. This is given in the article 'Nucleus' in the *Encyclopaedia Britannica* and is available as a wall chart published by the General Electric Company of America.

For the measurement of the electron charge see for example J. H. Saunders, *The Fundamental Atomic Constants,* Oxford University Press.

Chapter 12. For a concise account of rolling friction see P. Freeman, *Lubrication and Friction,* Pitman.

An account of certain properties of the eye, including adaptation, is given in the article 'Vision' in the *Encyclopaedia Britannica.* For a much fuller account see *Light Colour and Vision* by Yves Le Grand, the English translation of which is published by Chapman and Hall.

APPENDIX

Chapter 13. The extraordinary values derived by students for the conversion factor between the heat units are reported by Professor H. Wright Baker in the *Bulletin of Mechanical Engineering Education* No. 13.

Chapter 15. The experiments of Zahn and Spees are described in Volume 53 of the *Physical Review*.

For a thorough treatment of photometry see J. W. T. Walsh, *Photometry*, Constable.

INDEX

Absolute measurements in electricity, 114
Absolute temperature, 35, 36, 40
Absolute zero, 29, 32, 34, 37, 38
Absorption of radiation, 84
Action and reaction, 2, 3, 7
Acuity of vision, 92, 105
Adiabatic changes, 46, 47, 50-52, 104, 121
Algebraic problems in physics examination, 104-106
Alpha rays, 81-85
Ammeters, 71, 72, 105
Angle of contact, 8, 9, 111
Anharmonic oscillations, 22
Annihilation of electron-positron pair, 16, 62
Atomic bomb, 16, 83
Atomic physics, 115, 116, 119
Atomic structure, 85-89

Beta rays, 81-84
"Black-outs", 7
Bohr's theory of hydrogen spectrum, 86-89
Boiling of water, 110
Brownian motion, 123
Bulk modulus, 49, 95

Cadmium cell, 114
Calibration of electrical meters, 71, 72, 114
Calorie, 99, 113

Calorimetry, 113
Camera, 105
Candela, 114
Carnot cycle, 33, 34, 51
Carnot's theorem, 33, 52
Centre of mass, 11, 12
Centrifugal force, 3-6
Charles's law, 38-40
Circular motion, 3-7
Collisions: elastic, inelastic and superelastic, 23, 24, 106
Conductivity, thermal, 53-55, 58, 109
Conservation laws, 14-16
Convection, 55-58, 104
Cosmic rays, 60-62
Couples, effects of, 7, 8
Criticism, 122

Definitions, 97-103, 121
Deformation of bodies, 7
Depth of focus, 105
Discharge tube 115
Dogmatic teaching, 123
Doppler effect, 107, 108
"Dynamic equilibrium", 42
Dynamics, definition, 2
———, teaching, 98, 112, 121

Efficiency of ideal heat engine, 33, 34
Einstein's mass-energy relationship, 15-17, 122
Elastic hysteresis, 45, 91, 92

INDEX

Elasticity, 20
Electrical measurements, 114, 115
Electromagnetic theory, 68, 86, 89
Electromotive force (E.M.F.), 65-68
Electron capture, 61, 83
Electron charge, 89, 124
Electron, e/m, 90, 116
Electron-positron pair, 16, 62
Electron, wave properties, 90
Electroscope, 92-93
e/m for charged particles, 116
Emissivity, 56
Energy, chemical, 17
Energy, kinetic, 13
Energy, molecular, 20, 24, 117
Energy, nuclear, 16-17
Entropy, 49-52
Equations of state, 24-29, 39, 40, 43
Equilibrium, mechanical, 2
——, thermodynamic, 42
—— vapour pressure, 101
Equipartition, 117
Evaporation, 57, 101
Expansion coefficients, 22, 38
Extrapolation of laws, 32, 37
Eye, human, 92

"Fall-out", 83
Fission, 82
Fixed points, 31, 33
Force, 1, 2, 7, 20
Functions of state, 43

Gamma for gases, 116-118
Gamma-rays, 60-62, 81, 83, 84

Gas laws, 36-41
Gram-molecular volume, 110
Gravitation, 1, 10-12
Gravitational acceleration 'g', 5, 12, 98, 109
Gravitational red-shift, 108
Gauss, 77-79
Geometrical optics, 105, 119
Glass, 110

Heat engines, 33, 34, 51
Heat, nature of, 36, 43, 113
Heat radiation, 59, 60
Heat transfer, 53-58, 104
Heat transfer coefficient, 56
Heat units, 66, 99, 125
Hooke's law, 20, 21

Ice, growth on ponds, 58
Ideal gas, 30, 32, 35, 39, 40, 48, 107, 110
Increments, signs of, 95
Indigo, 111
Induced E.M.F., 67, 68, 73, 80
Inductance, 73, 74
Infra-red, 59, 60
Insulators for electrostatic experiments, 93
Intermolecular forces, 7-9, 18-29
Internal energy, 36, 43
Ionization by electron impact, 109, 115, 116
Irreversible changes, 44-52

Joule expansion experiment, 47
Joule-Kelvin effect, 47

INDEX

Kelvin temperature scale, 32-35
Kinetic energy, 13-15
—— theory, 22-30

Latent heat and surface energy, 20
"Least distance of distinct vision", 111
Leclanche cells, 111
Light, 59, 60
Linearity and proportionality, 56
Liquid metals, 55, 110

Magnetic circuit, 74
—— effect of electric current, 77
—— field intensity, 77, 78
—— induction, 78, 79
Mass, conservation, 14-15
—— definition, 97
—— variation, 1, 14
—— -energy relationship, 15-17, 122
—— spectrometer, 115, 116
Maxwell's thermodynamic relations, 28
Mechanical equivalent of heat, 113
Mercury thermometers, 32, 35, 36, 37, 40
Millikan's oil-drop experiment, 89
M.K.S. electrical units, 79
"Modern physics", 115
Molecules, properties of, 18-20, 22-24
Momentum, 1, 14, 106
Muon, 62

Neutrino, 15
Neutron, 82, 83
Newton's law of cooling, 56, 57
—— law of gravitation, 10
—— laws of motion, 1-5, 8
Non-ohmic materials, 69, 70, 73, 74
Nuclear energy, 16-17
Null measurements, 72, 115

Oersted, 77, 79
Ohm's law, 68-75
Oil-drop experiment, 89
Orbital motion, 4, 11, 12, 86

Partial differentiation, 96
Particles, 10, 11
Pendulum, 106
Permeability, 74, 76, 78, 79
Permittivity, 76, 79
Phonons, 55
Photoelectric cells, 60
Photoelectric emission, 94, 95
Photometry, 113
Photons, 15
Pin-hole, 109
Planck's constant, 60
Planets, 1, 10
Porous plug experiment, 47
Positron, 16, 62, 83, 84, 116, 124
Potential difference (P.D.), 65-68
Potentiometer, 70, 72, 94, 114, 115
Pressure, 26, 42
Proportionality and linearity, 56

INDEX

Protective clothing, 85
Proton, 82-83
Pupillary reflex, 92
Purkinje effect, 59

Quantum, 60, 94
Quantum theory, 86, 89, 117

Radiation hazards, 85
Radioactivity, 81-84
Relativity, 14, 107
Reluctance, 74
Resistance, 68-70
Resistivity, 100
Resistors, four-terminal, 93
Rest-mass, 14
Retina, 92
Reversible changes, 44-47, 50-52
R.M.S. values, 73
Rolling friction, 91, 124

Satellites, 1, 6, 10, 11
Semiconductors, 94
Shear, 7, 8
Sound Waves, 44, 47
Spacemen, 6
Specific heats of gases, 24, 116
—— —— of solids, theory, 118
—— ——, variation with temperature, 36, 40, 99
Spectra, 59, 62, 87
Spherical symmetry, 10, 12
Spring balance, 5
Statics, 2

Sublimation, 101
Surface tension, 7-9, 21
Syllabuses, 112, 122, 123

Temperature, 30, 31-41
Tension, 7
Tides, 11
Triple point of water, 32
Tube method for measuring conductivity, 58

Ultraviolet, 59
Unit cubes, 54, 100, 103

Van der Waals, equation of state, 26-29
Viscosity, 110
"Voltage", 65, 68
Voltmeters, 68, 71, 72

Weight, 2, 5-7, 98
Weightlessness, 5-7
Weston cadmium cell, 114
Wheatstone bridge, 72, 93, 115
Wiedemann-Franz-Lorenz law, 54, 109
Windows, conduction of heat through, 58, 104, 106

X-rays, 60-62, 83

Zahn and Spees, 116, 125
Zero point energy, 29